Applied Ethno–botany

Applied Ethno–botany

Dhiraj Acharya

RANDOM

RANDOM PUBLICATIONS

NEW DELHI (INDIA)

Applied Ethno-botany

ISBN 978-93-5111-370-6

Published in 2014 in India by

RANDOM PUBLICATIONS

4376-A/4B, Gali Murari Lal, Ansari Road
New Delhi-110 002
Phone : +91-11-43580356, +91-11-23289044
e-mail: randomexports@gmail.com, sales@randompublications.com, info@randompublications.com

Reprinted 2023

Type Setting by: Friends Media, Delhi-110089
Digitally Printed at: Replika Press Pvt. Ltd.

Preface

Ethnobotany has been defined as the discipline concerned with the interactions between people and plants. There are many aspects of Ethnobotany, including the ways that people name and classify plants, the values placed on them, their uses and their management. It reaches across the natural and social sciences. Plants have always been of central significance to human welfare and always will be. Plants provide people with food, fuel and medicine, as well as materials for construction and the manufacture of crafts and many other products. Their chemical and genetic constituents are being increasingly explored for human benefit. They are major parts of the physical worlds which people inhabit and in which they construct their realities. They lie at the base of food-webs, supporting most other forms of life. They are essential elements of ecological systems on all geographical scales, helping provide us with equitable climates, fertile soils and reliable supplies of water.

The central role of plants in the everyday lives of rural people is obvious in developing countries, with the daily round of activities revolving around agriculture, the care of domestic animals, the gathering of fuelwood, the cooking and eating of largely plant-based food, the construction of buildings and fences, the use of herbal medicine, and so on. Ethnobotany can be applied for many practical purposes, among which McClatchey and his co-authors have listed land-use development, agriculture, forestry, cultural conservation, education and the development of the health food and herbal medicine industries. Our interest here is in Applied Ethnobotany, which we define as Ethnobotany applied to conservation and sustainable development. Applied Ethnobotany draws on both personal (including traditional) and scientific forms of knowledge, allowing comparisons and integration for the benefits of conservation and sustainable development. In the past, ethnobotanical studies have all too often been just academic exercises or have served only external interests, with the results benefiting neither local people nor conservation. Our approach is cross-disciplinary, participatory, and geared towards local problem-solving. The betterment of rural livelihoods is a complex task and it is not suggested that Applied Ethnobotany is a universal panacea. Natural resource issues are only part of wider livelihood issues, and botanical issues only part of natural resource issues. There will always be uncertainties about how people will perceive their future options and the livelihood strategies that they will adopt. However, what is certain is that most rural people in developing countries will continue to be heavily reliant on local wild plants. Further, it is the poorest people, in particular, who will remain most dependent on resources of wild plants and who stand most to gain if these resources are managed in sustainable ways.

Applied Ethnobotany can therefore play a significant part in the battle to relieve poverty. The intention of the present paper is to provide information useful to those wishing to develop courses or programmes in Applied Ethnobotany. Many people have been consulted to produce the present synthesis, which certainly will require revision with time and as other experiences are incorporated. Each course or programme is expected to differ in detail, moulded as appropriate to the contexts of particular departments, universities and countries or regions. On the other hand, it is noteworthy that findings of the consultations undertaken for the present project show an unexpected degree of agreement about the contents of courses and programmes in Applied Ethnobotany and on how they should be taught. There appears to be an emerging consensus. Ethnobotany is sometimes seen as a 'soft' subject, but, actually, it can and should be intellectually demanding for both staff and students. As Janice Alcorn has commented "The ethnobotanical field worker observes the living system of which plant use is a part. Careful observation is not, however, an easy job. It requires looking at human activities and organisations from a plant's viewpoint and looking at plants and plant communities from a culturally informed as well as a scientific perspective.

This book has been designed as an introductory text which undertakes a through exploration the concepts and practices which define the subject. The book will prove very useful for students and teachers.

I would like to thank my team for standing beside me throughout my career and writing this book. My special thanks go to "Random Publications" who have published the book.

—*Dhiraj Acharya*

Contents

1

Introduction

Ethnobotany (*from "ethnology" - study of culture and "botany" - study of plants*) is the scientific study of the relationships that exist between people and plants.

Figure: *"Botanists at work" by Gerbrand van den Eeckhout, mid 17th century.*

Ethnobotanists aim to document, describe and explain complex relationships between cultures and (uses of) plants, focusing primarily on how plants are used, managed and perceived across human societies.

This includes use for food, clothing, currency, ritual, medicine, dye, construction, cosmetics and more.

History of Ethnobotany

Though the term "ethnobotany" was not coined until 1895 by the US botanist John William Harshberger, the history of the field begins long before that. In A.D. 77, the Greek surgeon Dioscorides published "*De Materia Medica*", which was a catalogue of about 600 plants in the Mediterranean. It also included information on how the Greeks used the plants, especially for medicinal purposes. This illustrated herbal contained information on how and when each plant was gathered, whether or not it was poisonous, its actual use, and whether or not it was edible (it even provided recipes). Dioscorides stressed the economic potential of plants. For generations, scholars learned from this herbal, but did not actually venture into the field until after the Middle Ages.

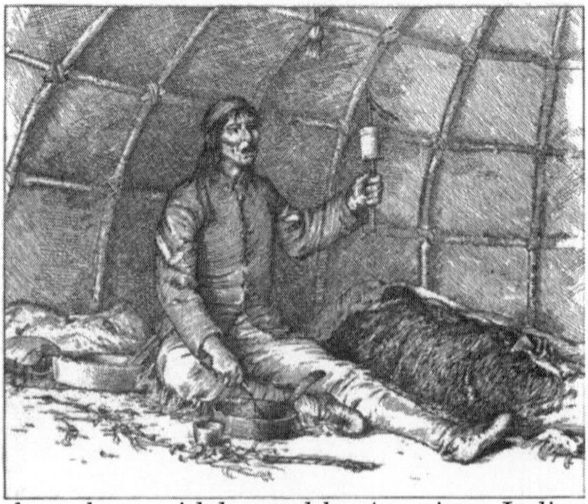

Figure: *Plants have been widely used by American Indian healers, such as this Ojibwa man.*

In 1542 Leonhart Fuchs, a Renaissance artist, led the way back into the field. His "*De Historia Stirpium*" catalogueed 400 plants native to Germany and Austria. John Ray (1686–1704) provided the first definition of "species" in his "*Historia Plantarum*": a species is a set of individuals who give rise through reproduction to new individuals similar to themselves.

In 1753 Carl Linnaeus wrote "*Species Plantarum*", which included information on about 5,900 plants. Linnaeus is famous for inventing the binomial method of nomenclature, in which all species get a two part name (genus, species).

The 19th century saw the peak of botanical exploration. Alexander von Humboldt collected data from the new world, and the James Cook's voyages brought back collections and information on plants from the South Pacific. At this time major botanical gardens were started, for instance the Royal Botanic Gardens, Kew. Edward Palmer collected artifacts and botanical specimens from people in the North American West (Great Basin) and Mexico from the 1860s to the 1890s.

Once enough data existed, the field of "aboriginal botany" was founded. Aboriginal botany is the study of all forms of the vegetable world which aboriginal peoples use for food, medicine, textiles, ornaments, etc.

The first individual to study the emic perspective of the plant world was a German physician working in Sarajevo at the end of 19th Century: Leopold Glueck. His published work on traditional medical uses of plants done by rural people in Bosnia (1896) has to be considered the first modern ethnobotanical work.

The term "ethnobotany" was first used by a botanist named John W. Harshberger in 1895 while he was teaching at the University of Pennsylvania. Although the term was not used until 1895, practical interests in ethnobotany go back to the beginning of civilization when people relied on plants for survival.

Other scholars analyzed uses of plants under an indigenous/local perspective in the 20th century: e.g. Matilda Coxe Stevenson, Zuni plants (1915); Frank Cushing, Zuni foods (1920); Keewaydinoquay Peschel, Anishinaabe fungi (1998), and the team approach of Wilfred Robbins, JP Harrington, and Barbara Freire-Marreco, Tewa pueblo plants (1916).

In the beginning, ethonobotanical specimens and studies were not very reliable and sometimes not helpful. This is because the botanists and the anthropologists did not collaborate their work. The botanists focused on identifying species and how the plants were used instead of concentrating upon how plants fit into people's lives. On the other hand, anthropologists were interested in the cultural role of plants and not the scientific aspect. Therefore, early ethnobotanical data does not really include both sides. In the early twentieth century, botanists and anthropologists finally collaborated and the collection of reliable, detailed data began.

Modern Ethnobotany

Beginning in the 20th century, the field of ethnobotany experienced a shift from the raw compilation of data to a greater methodological and conceptual reorientation. This is also the beginning of academic ethnobotany. The so called "father" of this discipline is Richard Evans Schultes even though he did not actually coin the term "Ethnobotany".

Today the field of ethnobotany requires a variety of skills: botanical training for the identification and preservation of plant specimens; anthropological training to understand the cultural concepts around the perception of plants; linguistic training, at least enough to transcribe local terms and understand native morphology, syntax, and semantics.

A great deal of information about the traditional uses of plants is still intact with the tribals. But the native healers are often reluctant to accurately share their knowledge to outsiders. Schultes actually apprenticed himself to an Amazonian shaman, which involves a long term commitment and genuine relationship.

In *Wind in the Blood: Mayan Healing & Chinese Medicine* by Garcia et al. the visiting acupuncturists were able to access levels of Mayan medicine that anthropologists could not because they had something to share in exchange. Cherokee medicine priest David Winston describes how his uncle would invent nonsense to satisfy visiting anthropologists.

Scientific Journals Covering Ethnobotanical Research

The Latin American and Caribbean Bulletin of Medicinal and Aromatic Plants (Abbreviated BLACPMA) or *Boletin latinoamericano y del caribe de plantas medicinales y aromaticas* (ISSN-0717 7917) is a bimonthly scientific publication directed to diverse professionals and technicians linked to the field of medicinal and aromatic plants.

It accepts papers related with the Bulletin's areas of interest, which are agronomy, anthropology and ethnobotany, industrial applications, botany, quality and normalization, ecology and biodiversity, economy and markets, pharmacology, phytochemistry, legislation, information and diffusion of events, courses, prizes, regulations, news, market questions, reports, biblioy, or any other material type that is important to publish.

See also the following peer-reviewed journals:

- *Journal of Ethnobiology*
- *Ethnobotany Research and Applications*
- *Journal of Ethnobiology and Ethnomedicine*
- *Journal of Ethnopharmacology*
- *Economic Botany*
- *Ethnobotany Research and Applications.*

Ethnobotanical Database

Some ethnobotanical database are as follows

- *Dr. Duke's Phytochemical and Ethnobotanical Databases*
- *Ethnobotanical Database of Bangladesh (EDB)*

Medicinal Plants

Medicinal plants are various plants used in herbalism and thought by some to have medicinal properties. Few plants or their phytochemical constituents have been proven to have medicinal effects by rigorous science or have been approved by regulatory agencies such as the United States Food and Drug Administration or European Food Safety Authority.

The articles in this category can be about traditional healing uses of plants, allopathic medicines derived from plants, and pharmacological research results about a plant.

Pharmacognosy is the study of medicines derived from natural sources, including plants. The American Society of Pharmacognosy as "the study of the physical, chemical, biochemical and biological properties of drugs, drug substances or potential drugs or drug substances of natural origin as well as the search for new drugs from natural sources."

Economic Plants

Our daily dependence on plant products of tropical origin is astounding. For instance, Latin America and Africa are major suppliers of coffee and cacao (from which we derive chocolate), while Asia produces most of our rice and natural rubber. Our lives are enriched by beautiful hardwoods, spices, essential oils and fruits. In addition, tropical countries export many fibres, gums, resins, dyes, and plant essences that we may never see directly, but which are widely used in medicine and industry. This section highlights some of these important plants.

The Tropics in World Trade

Plant products like those just mentioned are often referred to as "commodities" or "cash crops." Unlike many exports from the industrialized economies, commodities are usually exported in minimally processed states as raw materials. Whether tropical nations should continue to rely extensively on these exports to fuel their emerging economies is a hotly debated subject, with critics maintaining that overproduction depresses world prices of these materials and diverts arable land from food production for local markets. Regardless, patterns of trade in commodities are not likely to change significantly in the near future.

Trade vs. Environmental Concerns

As tropical nations seek to increase their share in the world marketplace, a key question is the best way to balance these strategies with the needs to conserve and manage remaining forested areas. Indiscriminate harvesting techniques and clearing large tracts for cultivation or ranching have been all too characteristic of the past.

The future will require more appropriate means of extracting plants or their products if we are not to lose the many thousands of other tropical species holding genetic "blueprints" important to our future. This will require strong international leadership on economic and environmental fronts and, for all citizens of the world, a willingness to rethink our use of the Earth's resources.

Native Origins of Economic Plants

Plants listed below are native to these regions. Many are now grown in other areas of the tropics also.

Essential Oils

Essential oils, or volatile oils, are found in many different plants. These oils are different from fatty oils because they evaporate or volatilize on contact with the air and they possess a pleasant taste and strong aromatic odor. They are readily removed from plant tissues without any change in composition. Essential oils are very complex in their chemical nature. The two main groups are the hydrocarbon terpenes and the oxygenated and sulphured oils.

These oils do not have any obvious physiological significance for the plant. They may represent byproducts or metabolism rather than foods. The characteristic flavour and aroma that they impart are probably to some advantage in attracting insects and other animals, which play a role in pollination or in the dispersal of the fruits and seeds. When in high concentration, these same odors may serve to repel enemies of the plants.

The oils may also have some antiseptic and bactericidal value. There is some evidence that they play an even more vital role as hydrogen donors in oxidoreduction reactions, as potential sources of energy, or in affecting transpiration and other physiological processes (Hill 1952).

All the distinctly aromatic plants contain essential oils. They occur in over 60 families and are especially typical of the Lauraceae, Myrtaceae, Umbelliferae, Labiatae and Compositae. The quantity of oil varies from a very small amount to as much as 1-2 percent. The oils are secreted by internal glands or in hair like structures. Sometimes, as in wintergreen and mustard, the oil is not present in the plant but develops only as the result of chemical action when the ground-up plant tissue is extracted with water. Almost any organ of a plant may be the source of the oil. Examples are flowers (rose), leaves (mint), fruits (lemon), bark (cinnamon), wood (cedar), root (ginger) or seeds (cardamom), and many resinous exudations as well.

These oils are extracted from the plant tissues in different ways depending on the quantity and stability of the compound. Three principal methods are: expression, distillation and extraction by solvents.

Perfumes

The history of civilization is directly connected with that of perfumes. Perfumes have been in widespread use since the earliest recorded times. The Egyptians and ancient Hebrews used them for both personal and religious purposes. They played an important role in the life of the Romans and Greeks, reaching such a high degree of specialization with the Greeks that a special perfume was required for each part of the body. Later Catherine de' Medici knew as much about perfumes as she did about poisons. In the time of Queen Elizabeth a gift of rare perfumes was a definite way to win the royal favour, while the court of Louis XIV at Versailles had a particular perfume for each day of the year, the preparation of which was supervised by the king himself.

In those days perfumes were of hygienic as well as aesthetic value for they acted as true antiseptics and deodorants and masked offensive odors at a time when bathing was infrequent. Perfumes have continued to be in great demand to the present day. The consumption of the natural products has gradually increased in spite of the many synthetic substitutes that chemists have placed on the market. Synthetics are not as long lasting as those obtained directly from the plants.

The most valuable perfumes are combinations of several essential oils. Frangipani, for example, contains sandalwood, sage, neroli, orris root, and musk, while one of the formulas for Eau de Cologne, which dates from 1709, calls for neroli, rosemary, lemon and bergamot dissolved in pure alcohol and aged. "The expert perfumer must be able to blend the several oils at his command as an orchestra leader combines the various instruments into a perfect whole" (Hill 1952).

Perfumes also contain fixatives, which are substances that are less volatile than the oils and which delay and so equalize evaporation. These may be of plant or animal origin. Musk, ambergris, and civet are frequently used for this purpose. Balsams and oleoresins, such as benzoin, styrax, and oak moss; essential oils with a low rate of evaporation like orris, patchouli, elary sage, and sandalwood; and various synthetic materials are also used.

Perfume plants are cultivated for the most part in areas bordering on the Mediterranean Sea and the Indian Ocean. Most of the natural perfumes are made in southern France in the region around Grasse and Cannes near the French Riviera. Here garden flowers are cultivated on a large scale, and from 10-12 billion pounds were being gathered annually by the mid 1950's. These included over 5 million pounds of orange blossoms, over 4 million pounds of roses, 440 thousand pounds of jasmine and 330 thousand pounds of violets. Large quantities of tuberoses, cassie, jonquils, thyme, rosemary, lavender and geraniums are grown and many other fragrant species to a lesser degree. Flowers are also grown for the perfume industry to some extent in Reunion, North Africa, England and various European, Pacific and Asiatic areas. When supplies were reduced during World War II, the United States developed substitutes and initiated or increased the cultivation of several essential oil plants in Central America. Of the 75 essential oils regularly used in the industry only eight are normally produced in the Western Hemisphere, and only oil of petitgrain is of much importance.

Perfume Oils

Some of the more important essential oils used in the manufacture of perfumes are as follows:

Otto of Roses

This is valuable oil that is also called Attar of Roses. It has been one of the most favourite perfumes either in combination with other oils or alone. Bulgaria supplied most the commercial supply in the 20th Century. The damask rose, *Rosa damascena,* was the main source. By the mid 1900's over 12,000 acres on the southern slopes of the Balkans were devoted to its cultivation.

The harvest period covers about three weeks during May-June. Flowers are picked in the early morning just as they are opening and are distilled immediately. In the beginning of this industry peasant farmers utilized their own primitive stills, but this gave way to larger modern distilleries.

The oil is colourless at first but gradually turns a yellowish or greenish colour. More than 20,000 lbs of the flowers are required to make one pound of the essence, which was valued at $200.00 in 1952. Very little pure Otto reaches the markets because it is almost always diluted with geranium or palmarosa oil or geraniol, which also have a rose like odor. Otto of Roses is also manufactured in France, Italy, North Africa, Asia Minor and India.

In France the cabbage rose, *R. centifolia* was used and the perfume was obtained both by hot and cold enfleurage as well as by distillation. Large quantities of rose water are also made. This consists mainly of the water left after distillation, which still contains some of the essence. Dissolving a small amount of Otto in water sometimes makes it.

Geranium

Pelargonium spp. leaves yield an essential oil after distillation. Geranium oil is widely used as an adulterant of or a substitute for Otto of Roses in making perfumes and soap. *Pelargonium graveolens* is most frequently grown especially in Algeria and Reunion and to a lesser extent in southern France and Spain. Cultural experiments have been made in Florida, Texas and California with *P. odoratissimum*, the rose geranium. The plants are easy to propagate from slips and are productive for 5-6 years after reaching maturity. They must be grown in minimum frost areas. A good grade of oil is obtained from the leaves of this species.

Ylang-ylang

A very valuable oil in the perfume industry, it is added to almost every perfume. The name translates as "flower of flowers." The ylang-ylang tree is an Eastern Asiatic species, *Canaya odorata*. Its yellowish-green, bell shaped flowers have an exceedingly delicate and evanescent fragrance. The oil is also known as Canaga Oil and is derived by simple distillation or extraction from the petals of fully opened blossoms. Most production was originally in The Philippines, but later French colonies in the Indian Ocean dominated the cultivation. This tree grows wild or cultivated in various parts of Southern Asia and the East Indies. The oil first arrived in Europe around 1864 and since that time it has been in great demand despite its high cost.

Cassie or Acacia

Flowers of the sweet acacia, *Acacia farnesiana*, yield an essential oil that is almost as valuable as Ylang-Ylang or Otto of Roses. It is a thorny small tree of the West Indies, but has spread to many tropical and subtropical areas. It is extensively cultivated in southern France, Algeria, Egypt, Syria and India as a source of perfume. The oil is removed from the petals by maceration with cocoa butter or coconut oil, or by extraction. It is similar to the odor of violets and is widely used for sachets, powders and pomades.

Neroli

This oil, obtained from orange blossoms, is extensively used in blends and for mixing with synthetic perfumes. True oil of neroli, or Neroli Bigarade, is distilled from flowers of the bitter orange, *Citrus aurantium*. Neroli Portugal is from the sweet orange, *Citrus sinensis*. Leading production has been in southern France, surrounding Mediterranean areas and in the West Indies, particularly in Haiti.

Other essential oils are also derived from Oranges and used in making perfumes. The leaves and twigs and sometimes immature fruits, supply Petitgrain Oil. This adds a pleasant bouquet to scents, cosmetics and soap. Paraguay has been the main producer. Bitter or sweet oranges are used and the oil is extracted by distillation. Oil of Orange is obtained by expressing the ripe orange peel, but it results in an inferior grade.

Bergamot

This is greenish oil that is expressed from the rind of the bergamot (*Citrus aurantium subsp. bergamia*). It has a soft sweet odor and has been widely used in the United States for adding scent to toilet soaps and in mixed perfumes. Italy and Sicily have been the chief exporters.

Orris

Rhizomes of *Iris pallida, I. fluorentina* and allied species contain an essential oil that has the odor of violets. Tincture of Orrisroot has been used to adulterate pure extract of violets and the powdered root is the basis of violet powder. Cultivation is in Southern Europe, Iran and northern India. Italian orrisroot is superior to other sources. The rhizomes are peeled and dried in the sun, and the odor gradually develops. Orris is also used as a flavouring substance.

Calamus

The roots of Calamus are the sweet and aromatic rhizomes of the sweet flag, *Acorus calamus*. It is a common plant of marshy ground in Europe, Asia and America. In powdered form Calamus is used for sachet and toilet powders, while the distilled oil is used in making perfumes. It has also been used for medicinal and flavouring purposes. The candied root was once a popular confection.

Grass Oils

Several important essential oils are derived from grasses and used in the perfume industry. The genus *Cymbopogon* (formerly *Andropogon*) is especially rich in perfume species.

Oil of Citronella

The oil is distilled from the leaves of *Cymbopogon nardus*. Java and Sri-Lanka have produced most of the world supply. The pale-yellow oil is inexpensive and used for making soaps and perfumes and as an insect repellent. The oil contains 80-90 percent geraniol, and is therefore an important substitute for Otto of Roses. Citronella was also introduced into Central America where a considerable industry developed in Guatemala and Honduras with over 4,500 acres under cultivation by 1952. The crop is harvested by hand. Cutting stimulates growth and a new crop is available in three months. The oil is extracted by steam distillation.

Lemon-grass Oil

Leaves of *Cymbopogon citratus* yield reddish-yellow oil with a strong odor and taste of lemons upon distillation. There is a very high content of citral in the leaves (70-80%). It is used in soaps and medicine. Citral is extensively used in perfumes, bath salts, cosmetics and toilet soaps and as a food flavouring. It is also the source of the aromatic substances known as ionones, which have many uses. One of the ionones is required in the synthesis of Vitamin A; another is the raw material for synthetic violet.

Lemon grass is common in the eastern tropics and is cultivated in Sri-Lanka, East Africa, India, the Congo and Madagascar. It was successfully introduced in the Western Hemisphere where large quantities of the oil were exported from Haiti, Brazil, Guatemala, El Salvador and other Neotropical areas.

Palmarosa / Ginger-grass Oils

These are nearly identical oils that have been used as adulterants of Otto of Roses. They contain a high amount of Geraniol. *Cymbopogon martinii* and related species are cultivated in India as a source and have been exported in large quantities.

Oil of Vetiver

The roots and rhizomes of the khuskhus plant, *Vetiveria zizanioides*, supply this oil. It is a native of Bengal and India, but has been grown throughout the tropics and subtropics. The roots have a very sweet scent and are made into mats, fans, screens, awnings, sunshades, baskets, sachet bags and pillows. The leaves are odorless. The plant was introduced into the West Indies and Louisiana and is widely used as an ornamental plant. It has escaped cultivation and become naturalized in many areas. On distillation the roots yield oil that is similar to citronella and which has been used for making expensive perfumes, soaps and in medicine. It is one of the best fixatives.

Oil of Bay

Leaves of *Pimenta racemosa* yield oil on distillation, which is used in perfumery and in the preparation of bay rum. The plant is native to the West Indies where the industry is located. Originally the leaves were distilled in rum and water, but now the oil is dissolved in alcohol and various aromatic substances are mixed in. Bay rum has soothing and antiseptic properties.

Lavender

Lavender perfumes are very old and were used by the Romans in their baths. It is still one of the most important scents. The true lavender plant, *Lavandula officinalis*, is native to Southern Europe, where it occurs on dry, barren soil. It is a low shrub with terminal spikes of very fragrant bluish flowers. There are many horticultural forms and hybrids occur. Lavender is grown in southern France at altitudes of 1500-1800 ft. Large amounts are also raised in England. Lavender has a clean odor and the dried flowers are used in sachets and for scenting chests and drawers. The oil is important in the manufacture of Eau de Cologne and other perfumes and is also used in soaps, cosmetics and medicine as a mild stimulant. Lavender water, a mixture of the oil in water and alcohol, is popular in England (Yardley brand).

Spike Lavender

This lavender, *Lavandula latifolia*, is coarser and yields an inferior grade of oil. It can be grown at lower altitudes than true lavender and is extensively cultivated in France and Spain. It is used in perfumes and cosmetics and to flavour meat sauces known as aspic.

Violet

One of the most popular perfumes is made from violets. Blue and pur0le double varieties of *Viola odorata,* native to Europe, are grown mainly in the vicinity of Nice. Solvents or maceration with hot fats extracts the oil. It occurs in such minute amounts that 15 tons of flowers are required to obtain only one pound of oil. Genuine violet perfume is rare and expensive, and it has been almost entirely replaced by synthetic products derived from ionone.

Jasmine

A highly esteemed perfume, jasmine is cultivated in southern France and surrounding areas. The main source is *Jasminum officinarum* var. *grandiflorum*, which is usually grafted on a less desirable variety. The flowers are picked as soon as they are open and the oil is extracted by enfleurage.

Carnation

There are thousands of horticultural varieties of carnation all derived from *Dianthus caryophyllus*, a species of southern Europe, Northern Africa and tropical Asia. The most conspicuous blossoms and colour give the least odor. Therefore, for the perfume industry varieties with less conspicuous blossoms are used. Solvents extract the oil. Synthetic carnation oil predominates in world markets.

Rosemary

Rosemary, *Rosmarinus officinalis*, is a native of the Mediterranean region. It has long been a favoured sweet-scented plant and has been important in the folklore of many countries. It is one of the least expensive and most refreshing odors. The plant is a small evergreen shrub that is cultivated in Europe and the United States. The oil is extracted by distillation of the leaves and fresh flowering tops or by extraction. It is used in Eau de Cologne, toilet soap and medicine. The leaves are valuable as a spice.

Hyacinth

Hyacinthus orientalis is native to Western Asia and Asia Minor. It was introduced into Europe in the 16th Century where it was grown as an ornamental plant especially in The Netherlands. It is also a familiar species in North America. Hyacinths are grown for perfume in southern France. The odor is heavy, sweet and quite overpowering. Solvents are used to obtain the oil, which is generally greatly diluted.

Oak Moss

Oak Moss, also called Mousse de Chene, is a valuable addition to the raw materials of the perfume industry. It comprises various lichens that grow on the bark of trees. The main sources are European species of *Ramelina* and *Evernia*, particularly *R. calicaris, E. furfuracea* and *E. prunastri*. These lichens contain oleoresins that are extracted by means of solvents. After they have been collected, the lichens are dried. Then the perfume develops in storage. Oak moss not only has a heavy, penetrating odor and blends well, but also has a high fixative value. It is an essential element in lavender perfumes and soap and in the better grades of cosmetics.

Linaloe or Bois de Rose

Several sources of this very aromatic substance occur. Mexican linloe is distilled from wood chips of two species, *Bursera penicillata* and B. glabrifolia. Cayenne linaloe or Bois de Rose is derived from *Aniba panurensis* of the Guianas, while Brazilian Bois de Rose is from *A. rosacodora var. amazonia*, a tree in the lower Amazon basin. The product is widely used in perfumes, soaps, and cosmetics and for flavouring of foods and beverages.

Sandalwood

The oil is obtained by distillation from the wood of *Santalum album* and related species. The tree grows wild in India and other parts of Southeastern Asia and is cultivated in many other areas. The oil is used throughout the Orient as a perfume and also in medicine. It is an excellent fixative and is used in blends. The sweet-scented wood is made into chests and boxes. Demand for sandalwood has been very great, resulting in the eradication of the species in many areas. Several substitutes have been used.

Patchouli

This oil is obtained from the flesh leaves and young buds of Pogostemon cablin. The plant is a small shrub that grows wild in Southeastern Asia and is cultivated in China. The leaves are fermented in piles and are then distilled. The dark-brown oil has a powerful odor, resembling that of sandalwood. It is one of the best fixatives for heavy perfumes. It is also used in soaps, tobacco and hair tonics. It imparts the characteristic odor to cashmere shawls, which are shipped in patchouli-scented containers.

Champaca

This oil constitutes one of the most famous perfumes of India and other oriental countries. It is obtained from *Michelia champaca*, a large tree of the eastern tropics. The conspicuous yellow flowers are very fragrant and are frequently worn by the natives. The oil is removed from the flowers by maceration or extraction and rivals ylang-ylang in its delicious fragrance.

Misc Perfume Sources

Some of the gum resins, mainly frankincense and myrrh, have been used in perfumery for thousands of years.

Other garden flowers that are cultivated for their perfume include Heliotrope (*Heliotropium arborescensj*), Lily of the Valley (*Convallaria majalis*), Jonquil (*Narcissus jonquilla*), Mignonette (*Reseda odorata*), Narcissus (*Narcissus tazetta*), Clary Sage (*Salvia sclarea*) and Tuberrose (*Polianthes tuberosa*).

Oils from caraway, anise, cassia, cinnamon, clove, peppermint, lemon, thyme wintergreen and zedoary have also been used in the perfume industry, but these are discussed under other categories.

2

Forest Resources

Distribution of Forest Lands

Forests occur in some parts at least on all continents. Although knowledge concerning the exact nature and distribution of these forests is still incomplete, it is possible to estimate the amount of forest land in the world in 2000 as about six billion acres, or 18 percent of the earth's surface. This area is progressively decreasing as the human population soars.

These forests are distributed on the different continents as 25% in Asia, 24% in South America, 18% in North America, 8 percent in Africa, 9% in Europe and only 3% in Australia and Oceania. Forests were, of course, originally much more abundant than they are at the present time.

The softwoods or conifers comprise 34 percent of the total forest area occurring in pure stands or mixed with hardwoods. They are especially abundant in the colder climates, about 90% occurring in the north temperate zone. When they are found in warmer regions, conifers are restricted to the higher altitudes unless they have been imported as exotic species (e.g., Monterrey Pine). For hardwoods a distinction is usually made between temperate and tropical species.

The former occur 16% and the latter 35% of the total forest area. As in the case of conifers, most of the temperate hardwoods (ca. 85%) are located in the north temperate zone. It is interesting that three-quarters of the world's population lives in this area and consumes nearly 90% of all the wood used. It is of great economic importance that both softwoods necessary for general construction and hardwoods have been readily available throughout this area. As a result of the depletion of native forests, attention has been turned to the tropical forests, which continue to be decimated by clear cutting.

Forests of North & Central America

The forests of North America occupied about 25 percent of the land area by the beginning of the 21st Century. Conifers comprised about 72 percent, temperate hardwoods about 20 percent and tropical hardwoods about 8 percent of the forests. The northern portion of North America (Alaska, Canada and Newfoundland) is primarily coniferous, with 90 percent of the softwoods and 7 percent of temperature hardwoods.

The United States had about 62 percent conifers and 38 percent temperate hardwoods. Mexico had about 47 percent conifers, 34 percent temperate hardwoods and 19 percent tropical hardwoods. In Central America tropical hardwoods accounted for 75 percent and conifers 25 percent of the forests.

The West Indies had an even higher percentage of tropical hardwoods (ca. 85 percent), with the remainder in conifers. The forest area present by the 21st Century was only about half that of the original forests that existed at the time of the first European settlements that began in the 16th Century.

Types of Forests In North America

Forests in North America are found in about six well-defined areas: (1) northern coniferous forest, (2) eastern deciduous forest, (3) southeastern coniferous forest, (4) Rocky Mountain forests, (5) Pacific Coast forests and (6) tropical and subtropical forests.

Northern Coniferous Forest

This is a predominantly evergreen forest that extends across the continent from Newfoundland and Labrador to the lower Hudson Bay region and Alaska, south of the treeless arctic tundra. The principal trees are white and black spruce, balsam fir and larch, with some paper birch, aspen and balsam poplar. From Nova Scotia and northern new England to Minnesota and southward along the summits of the Appalachians, there is a transitional region between the coniferous forest and the eastern deciduous forest, with species of both these areas mixed.

The prominent trees of this Northern Hardwood Region, include white pine, red spruce, white cedar, beech, sugar maple, hemlock, yellow birch and some red pine and jack pine.

Eastern Deciduous Forest

This is one of the oldest forests on the North American continent that covers most of the eastern and central portions of the United States. It flourishes in the lower Ohio valley and on the slopes of the southern Appalachians, and it extends as far north as Ontario and southern Quebec. Prominent trees are oaks, hickories, tulip, chestnut, black walnut, ash, basswood and formerly American elm. At the northern limits the beech and maple become prominent mixed with different conifers. Towards its southern and southwestern limits the oaks and hickories occur with many of the pines that are typically found in the southeastern coniferous forest. Westward the deciduous forest gradually is confined to the river valleys of the prairie region.

Southeastern Coniferous Forest

This forest occurs along the sandy Atlantic coastal plain from Texas to Virginia. Different species of pines, mainly shortleaf, longleaf, loblolly and slash pines, occur on the uplands; in lower ground there occurs the tupelo, live oak, red gum and magnolia. Southern white cedar and cypress are found in swamps.

Rocky Mountain Forest

This forest consists mainly of coniferous species. The area extends from northern British Columbia southward across the United States and Mexico and on into Central America. There are many differences in the nature of the forest as influenced by latitude and altitude. Western yellow pine is the most typical species. Others include lodgepole pine, white fir, Douglas fir, and western larch. Engelmann spruce and alpine fir are at the higher elevations. In northern Idaho and Montana a forest occurs that is similar to some of those found on the Pacific Coast, with western red cedar, western hemlock and western white pine being the main species.

Pacific Coast Forests

There are several distinct forest areas located within the Pacific Coast region. Along the coast from Alaska south to Washington, Sitka spruce is the main species. In southern British Columbia, the Puget Sound region and eastern Oregon and Washington as far west as the summits of the Cascade Range there is situated one of the most magnificent conifer forests in the world.

It can hardly be surpassed in its density and the size of the trees, which reach heights of 200-250 ft. and diametres of 8-15 ft. Mild winters are created by the nearness of the Pacific Ocean and there is a very high precipitation that can exceed 100 in. per year. Douglas fir is the dominant species and associated with it are western hemlock, western cedar, Sitka spruce and several species of fir.

Beginning in southwestern Oregon and extending past the San Francisco Bay area the forests along the Coast Range are dominated by redwoods.

Much of the original redwood forest had been cut by the end of the 20th Century and farmland was substituted.

East of the Cascade Range the forests merge with those of the Rocky Mountains. The main trees are western yellow pine, western larch, Douglas fir, lodgepole pine and some western white pine, Engelmann spruce and alpine fir. This forest also extends southward along the Sierra Nevada Mountains in California.

There western yellow pine, sugar pine, incense cedar, Douglas fir and white fir are common with red fir at the higher elevations. The Big Trees are found in isolated stands along the lower slopes of the Sierra Nevada.

Tropical Forest

Although subtropical conditions are found in south Florida, none of the trees are of commercial importance.

The true tropical forests begin in the lowlands of Mexico (south of El Mante in the east and Mazatlán in the west), on the eastern slopes of Central America and in the West Indies.

Most of the original forest in the latter two areas has been destroyed as a result of migratory agriculture and has been replaced by the dense, almost impenetrable tropical jungle.

The more important tropical species of North America will be discussed along with those of South America.

Major Woods of Temperate North America

There are more than 500 species of temperate woody plants in North America, with about 100 being of commercial importance. The principal species are discussed as follows:

Softwoods

Cedar: These trees all have a fragrant, light, soft wood that is even grained and decay resistant. The wood is used mainly in millwork and in the manufacture of woodenware, instruments, caskets, boats and various appliances. Commercial cedar includes seven different species. Port Orford Cedar, *Chamaecyparis lawsoniana,* is found in Oregon and California. It is the hardest of all the cedars and has some exceptionally good technical properties. It yields a strong, durable, heavy, stiff timber that takes a good polish. The wood is used for furniture, boats, millwork, matches, floors, interior finish and posts.

Alaska Cedar, *Chamaecyparis nootkatensis,* occurs from Alaska to Oregon along the Pacific Coast. The wood is light, stiff, hard and strong and is easy to work and durable. It is used for boats, shingles, fences and interior finish.

Southern White Cedar, *Chamaecyparis thyoides,* is a smaller tree found in the Atlantic and Gulf Coast states. The wood is useful for boats, poles, shingles, ties, posts and cooperage. During colonial times it was used to build single-story houses.

Incense Cedar, *Libocedrus decurrens,* of California, western Nevada and Oregon has a close-grained reddish wood of high value. It is used in building, for posts and piling and to a great extent for lead pencils.

Eastern Red Cedar, *Juniperus virginiana,* is one of the most common trees of the eastern United States. The wood is very durable and is particularly resistant to weathering. It is used primarily for fence posts, poles, crosstrees and railroad ties. The sapwood is white and the heartwood is rich red and very fragrant due to the presence of an essential oil.

The wood is soft with an even, fine grain; and it can be whittled easily. Panels, veneers chests and interior finish are made from eastern red cedar.

Northern White Cedar or Arbor Vitae, *Thuja occidentalis,* is a common tree in New England and the Great lakes area and adjacent Canada It is soft and easily worked and has been used mainly for canoe and boat building, woodenware, shingles, fence posts, railroad ties, poles, tanks and silos.

Western Red Cedar, *Thuja plicata,* occurs in the northern Rocky Mountains and Pacific Northwestern United States. It is the largest of the cedars that may reach a height of 150 ft. and a girth of 30 ft. The brownish-red, close, even-grained wood is soft but very durable. Most of this wood has been used for shingles. Other uses include poles, fences, cabinetwork, interior finish and cooperage. The coastal Amerindians used this species for their totem poles and war canoes, and also fabricated ropes and textiles from the bast fibres of the inner bark.

Cypress

The Bald Cypress, *Taxodium distichum*, is a tree of the swamps along the Atlantic Coast from Delaware to Texas and northeastern Mexico, and in the lower Mississippi valley. It is one of the strongest and heaviest of the softwoods. It may reach a height of 80-140 ft. and a girth of 5-12 ft. The "cypress knees" are a typical feature, which are conical outgrowths from the roots for the purpose of aeration.

The tree is deciduous and the wood is a rich red colour with a distinct grain. It is soft and coarse and is easy to work with. It has been used primarily in millwork for cabinet and interior finishes. It is also used for shingles, ties, posts, tanks, cisterns and other structures that are prone to decay, because the wood is very durable. Other products that have been manufactured from bald cypress are boats, cooperage and boxes.

Douglas Fir

Douglas fir, *Pseudotsuga taxifolia*, is one of the most economically valuable woods in North America. The tree covers a wide range in northwestern United States and western Canada. It attains its best development on the Pacific Coast from central California to British Columbia. The trees may reach a height of 200 ft. and a diametre at the base of 8-10 ft. The lowest branches are high on the trunk so that the trees have been used frequently for masts, flagpoles and spars. The size of the trunk also makes possible timbers of many lengths and sizes. The wood is resinous with a close, even, well-marked grain, and is of medium weight, strength, stiffness and toughness. It is very durable and, when well seasoned, will not warp.

Much Douglas fir is used in heavy construction. Other uses include railroad ties, paving blocks, piles, veneers, floors and millwork. The bark has been processed for use as a substitute for cork in making adhesives, plastics and explosives. It also possesses insecticidal properties.

Fir

True firs are of comparatively little economic importance. The wood is very light, soft and brittle and has been used mainly in the manufacture of crates and boxes. Fir is also used as a source of pulpwood and in millwork. The most important species are the balsam fir, *Abies balsamea*, of the northern coniferous forest and four western species, the lowland white fir, *A. grandis*, , the red fir, *A. magnifica*, the noble fir, *A. procera* and the white fir, *A. concolour*.

The last two species when young are of commercial importance as Christmas trees. Because of their dense stands in portions of the American western mountains, they are extremely valuable as watershed species.

Hemlock

Eastern Hemlock, *Tsuga canadensis,* is one of the most characteristic trees of the northeastern transition forest. It furnishes an inexpensive coarse lumber that is used primarily for framing timbers, sheathing, scantling, laths, rafters and other types of crude construction.

Western Hemlocks, *Tsuga heterophylla* & *T. mertensiana,* are larger trees that have a superior wood, which is heavier, stronger, stiffer and more adapted to heavy construction. Both species are high in tannin. These species are important pulpwood sources in the Pacific Northwest.

Larch

Eastern Larch or Tamarack, *Larix laricina,* and Western Larch, *L. occidentalis,* are the larch wood of commerce. The first species is found in the northeastern United States and across Canada, while the latter species is a larger tree that grows mainly in Idaho, Montana and Washington. Larch is very heavy, strong and tough softwood and is used mainly in furnishing heavy timbers for general building construction. Because larch is very durable it has been used for posts, poles, fences, railroad ties and paving blocks. The naturally curving lower parts of the trunk furnish ideal material for boat "Knees," ribs and other forms of ship timber (Hill 1952). Larch is also used to manufacture planning mill products, boxes and tanks.

Pine

Pines have always been of great importance commercially, and they constitute about half of the total lumber supply. The wood is obtained from different species that belong either to the soft pine or the hard pine category. In North America there are eight species of especial value. The soft or white pines have a straight-grained soft wood of mellow and uniform consistency that is relatively free of resin and is easy to work. It is used for rough carpentry, cooperage, cabinetwork, toys, boxes and crates.

The hard or yellow pines have a resinous, heavy, strong, hard and durable wood that finds great use in buildings, bridges, ships and other types of heavy construction. Because it is very durable, yellow pine is desirable for stairs, floors, planks and beams.

White Pines

Northern White Pine, *Pinus strobus*, is one of the most valuable timber trees. It was formerly used more than any other species, both for domestic consumption and for the export trade from America to Europe. By the 21st Century so much of the wood had been cut that it fell into sixth place in importance. The tree is found in the northeastern United States and adjacent Canada, and along the Appalachian Mountains to Georgia. Considerable stands still exist in Michigan, Wisconsin, Minnesota and a small area of northwestern Illinois. It is a conspicuous member of the forest flora in this region and trees may attain a height of 100-200 ft. and a diametre of 3-9 ft.

The wood is very light and easy to work as it is one of the softest of timbers. However, it is very durable. Houses that had been constructed of white pine in colonial days remain in good condition. The sapwood is white and the heartwood a pinkish brown, with a fine, even grain and lustrous surface. White pine has been used for building structures, doors, windows sashes, cabinetwork, boxes, and matches. Other species with similar wood and uses that have been called white pine in the lumber trade are Western White Pine, *Pinus monticola*, Sugar Pine, *P. lambertiana*, one very large and valuable timber tree of the West, and Lodgepole Pine, *P. contorta*.

Yellow Pines

Yellow pines are a heterogeneous group that is often classed together even though each species has a distinctive wood. Eastern species are very fast growers. The southern yellow or Longleaf Pine, *Pinus australis*, is one of the two most important timber trees of the United States. It occurs in the southeastern states from North Carolina to Texas. The wood has a fine, smooth, compact grain and is the heaviest, hardest, strongest, stiffest and toughest of the softwoods.

It is especially durable and able to bear great weights. Therefore, it main use is for beams, joists and other timbers for heavy construction, and for wharves, ships, bridges and railroad ties. Some yellow pine is used for boxes and millwork. The wood is very resinous and the tree is the main source of the naval-stores industry. It is also an important pulpwood. Associated with the yellow pine in the southern forests and often classified with it are the Shortleaf Pine, *Pinus echinata*, the Slash Pine, *P. caribaea*, and the Loblolly Pine, *P. taeda*.

Other important hard pines are the Western Yellow Pine, *Pinus ponderosa*, and Red Pine or Norway Pine, *P. resinosa*, of the Eastern United States. Red pine is of considerable economic importance in the Christmas Tree trade, the trees being grown in some quantity on farmland.

Redwood

Coastal Redwood, *Sequoia sempervirens*, (Photos) is restricted to the Pacific Coast area of central and northern California and southwestern Oregon. It is one of the tallest trees in the world that can reach a height of 200-300 ft. and a diametre of 8-122 ft. Only one other species, the Mountain Sequoia, *Sequoia gigantean*, (Photos) can exceed these dimensions and also attains a greater age estimated at 3,600 years. The wood of the Mountain Sequoia has little commercial value and has consequently escaped decimation by logging practices. However, Coastal Redwood has been widely exploited in the 20th Century and only the efforts of conservationists have made possible the preservation of a few stands of primeval forest.

The wood of Coastal Redwood is fine and straight-grained, strong, light and very soft. The sapwood is pale while the heartwood has a rich dull red colour. The cinnamon-brown bark, often one foot thick, is striking. Redwood does not warp or shrink readily and is very durable, especially after seasoning. It has been used for general construction, shingles, siding, tanks, coffins, silos, posts, water pipes, ties, furniture, cabinetwork and interior finish. The fibrous bark ahs been used as an insulating and stuffing material, and it yields a textile fibre for use with wool. The fine bark dust is used as a soil conditioner.

Spruce

Several species of spruce have a light, soft, compact, straight-grained wood that is stiff, strong, and easy to work and relatively free of resin. Spruce has increased in importance as white pine has become less available. The main uses are for pulpwood, light construction, boxes, millwork, and cooperage and as Christmas trees.

The wood is resonant and so is used for making the sounding boards of pianos and the bodies of violins and other instruments. The main species is the White Spruce, *Picea glauca*, one of the most characteristic trees of the great coniferous forest that extends from the northeastern United States to Labrador and across the continent to Alaska. Other eastern species are the Red Spruce, *Picea rubens*, and the Black Spruce, *Picea mariana*, both of which have a more restricted range.

The most important spruce species of Western North America is the Sitka Spruce or Tideland Spruce, *Picea sitchensis*. This large tree may attain a height of 200-300 ft. It occurs along the coast from Alaska to northern California. Sitka spruce has found use in the manufacture of oars, boats and other wood products that require a light, strong and elastic wood. It was once widely used in airplane construction. Engelmann Spruce, *Picea engelmannii*, occurs in the Rocky Mountain and Cascade Range region from Arizona and New Mexico to Canada, and has had limited economic importance.

Hardwoods

Ash

This is a strong, tough, elastic, stiff, hard and lightweight wood. It is light reddish brown, easy to split and difficult to nail. It often bears beautiful figures and may be highly polished. Other uses of ash wood are the manufacture of oars, bats, handles, tennis rackets, cues, rods, clothespins, toys, baskets and barrels. It may also be used for building carriages, boats, farm implements, furniture, cooperage and interior finish. The most widely used species is White Ash, *Fraxinus americana*, of the eastern deciduous forest in North America. Other prominent species are the Red Ash, *Fraxinus pennsylvanica*, and a variety the Green Ash, *Fraxinus pennsylvanica var. lanceoleata*; the Blue Ash, Fraxinus quadrangulata; the Black Ash, *Fraxinus nigra*, and the Oregon Ash, *Fraxinus oregona*.

Basswood

Also called Linden, *Tilia americana*, occurs mainly in the eastern deciduous forest of North America. The tree may reach a height of 80 ft. The wood is light coloured and straight grained with a smooth uniform texture. It is light, soft and weak and not very tough. However, because of its colour, even grain and ease of working it is widely used. It has been used to manufacture boxes and crates, millwork, woodenware, furniture, trunks, Venetian blinds, picture frames, carriage bodies, beehives, plywood, cooperage, pulp, charcoal and excelsior.

Beech

The Northern Beech, *Fagus grandifloia*, is typically found in the northeastern transition forest of North America. The wood is fine-grained in pinkish brown in colour. It is moderately strong, hard and heavy and has a wide range of uses. Beech has been used extensively to make boxes and crates because it does not impart any taste or odor. Flooring, interior finish, fixtures and furniture, tool handles, woodenware, clothespins, wagon stock, shoe lasts and ties are among the other products. The wood is also used for fuel, charcoal and wood-distillation.

Birch

This is a heavy, hard, strong and tough wood with a fine wavy grain that is often beautifully figured and able to take a high polish. It is frequently stained and finished to imitate mahogany or cherry. The Yellow Birch, *Betula lutea,* that ranges from the Lake States to New England and Georgia, and the Black Birch, *Betula lenta,* which has a more restricted range, furnishes most of the wood used in the manufacture of doors, furniture, window frames, floors and other forms of millwork. Other articles include clothespins, handles, shoe pegs and lasts, wheel hubs, woodenware, boxes baskets, dowels, yokes, veneers and spools.

White Birch or Paper Birch, *Betula papyrifera,* occurs further north on the continent of North America. The wood is very strong and elastic with a fine uniform texture. It is used mainly for plywood, spools, toothpicks, boxes, handles, dowels, bobbins and shoe lasts and pegs and in turnery. Especially the Amerindians have used the bark that peels off in typical layers to make canoes and fancy articles. Birch is also used for fuel and distillation.

Cherry

Wild Black Cherry, *Prunus serotina,* is the only one of several species in the same genus where the wood has commercial value. The tree occurs in the deciduous forests from Ontario to Florida and from the Dakotas to Texas. It is particularly abundant in the southern Appalachians. The wood has a fine, straight and close grain, and it is hard. The colour varies from light to dark red depending on age, and it is often stained before use. The beautiful grain and colour and the ease with which it can be worked make this cherry especially desirable for furniture, interior finishes and cabinetwork. The supply had become greatly reduced by the end of the 20th Century. Other uses have been the stands for scientific instruments, printer's supplies, and turnery. Medicinal properties are in the bark, and the ripe fruits are at times used to flavour liqueurs.

Chestnut

The American Chestnut, *Castanea dentata,* was one of the most important timber trees of the eastern United States. However, the accidental importation of chestnut blight disease from Europe practically eradicated the tree. Before the disease epidemic American Chestnut was very abundant and conspicuous because of its large size in the deciduous forests from Maine to Tennessee. It was especially plentiful in the southern Applachians. The wood was brown, soft and open grained and very durable and easy to work. It was used for millwork, caskets, furniture, musical instruments, woodenware, boxes, veneers and plywood cores. Its high durability made it excellent for poles, posts, piling, ties, fence rails, shingles and cooperage. It was also a source of pulpwood and a tanning material. The nuts were superior in quality to the European Chestnut. The tree continues to survive from suckers that arise from the roots, but they never reach a reproductive age.

Elm

Several elm species, especially the Rock Elm, *Ulmus thomasi*, and the ornamental White Elm, *Ulmus americana*, yielded valuable wood with a beautiful grain. The wood of the Rock Elm is strong, tough, hard, elastic, heavy, pliable and durable. It was the most important source of hubs, spokes, fellies and wheel rims. It was also used for agricultural implements and tool handles, butcher blocks, veneers and cooperage and in the manufacture of furniture, musical instruments, woodenware and baskets.

The White Elm has a softer, lighter and weaker wood that was nevertheless tough and fibrous and had been used for the same purposes. Dutch Elm Disease had by the end of the 20th Century eliminated over 98 percent of the White Elms. The tree survives only in small, mostly far northern pockets where there is some protection from the vectors of both a virus and the bacterium.

Hickory

Hickory trees are found in the eastern deciduous forest from Ontario to Minnesota, Florida and Mexico. The wood of all species is used to some extent, but the most important sources are the Shagbark Hickory, *Carya ovata*, the Mockernut, *Carya tomentosa*, and the Pignut, *Carya glabra*.

The wood is one of the toughest, hardest, heaviest and strongest and it is used where both strength and the ability to withstand shocks are required. It is a coarse straight-grained wood.

The sapwood is preferred to the heartwood. Hickory has been used for spokes, fellics, axles and other parts of wagons, and also for ax, pick and hammer handles, baseball bats, agricultural implements, shafts of golf clubs, pump rods and cooperage. It is the standard for fuel wood and for smoking meat.

Locust

Black Locust, *Robinia pseudoacacia,* is another tree of the eastern deciduous forest of North America. The wood is very heavy, strong, hard, durable and elastic. It has a coarse, open, crooked, compact grain and a smooth, satiny surface. It has been used mainly for insulator pins and brackets. It was also commonly used for tree nails, boat ribs, fence posts, ties, sills, wagon hubs and mine timbers. The wood is of more importance in Europe where it has been used for furniture and shipbuilding.

Honey Locust, *Gleditsia triacanthos,* has similar properties and uses. The leaflets are much smaller than the Black Locust and the flowers are aromatic. This species has long thick seedpods.

Maple

Maples constitute some of the most important woods in North America. The principal source of the commercial wood known as Hard Maple is the Rock Maple or Sugar Maple, *Acer saccharum.* This tree is a conspicuous species of the eastern deciduous and northeastern transition forests. It ranges from southeastern Canada to Georgia. Two other easier species, the Silver Maple, *Acer saccharinum,* and the Red Maple, *Acer rubrum,* and the Oregon Maple, *Acer macrophyllum,* and the Oregon Maple, *Acer macrophyllum* that occurs in Washington and Oregon, furnish soft maple wood that is of lesser importance.

Hard maple wood is tough, heavy, compact, strong and very hard. It is light brown in colour with dense even grains and a fine texture. It may be finely polished and is often beautifully grained and figured, as in the case of bird's eye maple and curly maple. These latter traits make it one of the best woods for furniture, veneers, flooring, interior finish and bowling alleys. It is also used for violins and other musical instruments, shoe lasts, rulers, tool handles, inlays, panels, keels of vessels, pianos, bowls, cooperage, charcoal, fuel and wood distillation products. The sap of the hard maples yields maple syrup and maple sugar.

Oaks

Oak has been regarded as the most important of all hardwoods. The timber has great commercial value, but the tree is also the largest and finest of the hardwood forests. The wood is tough, hard, resilient, durable and elastic. Its strength and ability to resist heavy strains make it invaluable in shipbuilding and other types of heavy construction. It also polishes easily and is unmatched for decorative work. There are more than 50 species of oaks in North America, about twenty of which are of commercial importance These belong either to the white ok or to the red oak group. Because distinguishing between the woods of the different species, they are regarded as either white or red oak.

White Oak Group

Lumber of white oaks is harder, stronger and more durable than of red oaks. It is used for used for building timbers, piling, railroad ties, machinery parts, agricultural implements, furniture, flooring, cabinetwork, interior finish an cooperage. It is interesting to note that oak barrels are the most satisfactory containers for wine, beer and alcoholic spirits. The most important species in this group include the White Oak, *Quercus alba*, the Bur Oak, *Quercus macrocarpa*, the Post Oak, *Quercus stellata*, the Chestnut Oak, *Quercus montana*, the Swamp Chestnut Oak, *Quercus prinus*, the Oregon Oak, *Quercus garryana*, and the Swamp White Oak, *Quercus bicolour*.

Red Oak Group

Lumber of the red oaks is softer, more open grained, more porous and less durable than that of the white oaks. The main species include the Red Oak, *Quercus borealis*, the Black Oak, *Quercus velutina*, the Scarlet Oak, *Quercus coccinea*, the Pin Oak, *Quercus palustris*, the Turkey Oak, *Quercus laevis*, the Willow Oak, *Quercus phellos*, the Texas Red Oak, *Quercus texana*, and the Shingle Oak, *Quercus imbricaria*.

Live Oak Group

There are several species of live oaks in North America, but the most important is *Quercus virginiana*, which as the hardest, strongest and toughest wood of all the oaks. It has been used in the construction of ships, wagons and farm tools. There have never been great quantities of this species available.

Osage Orange

Osage orange, *Maclura pomifera*, is a small tree native to the Gulf States but which has been cultivated in other areas. The wood is the heaviest, toughest, hardest and most durable of all the North American hardwoods. Only a small amount has ever been available, which was used mainly for fellies, tree nails, insulator pins, posts, stakes and woodenware. The Amerindians used this wood for their bows, and it is the source of a dyestuff.

Poplar

True poplars in North America are referred to under several different names. The most important is the Cottonwood, *Populus deltoids*, which is found in the Central and Eastern portions of the continent. It has a soft, light, even-grained fibrous wood that is easy to work with. Its main uses are as a substitute for basswood in the manufacture of boxes and excelsior and as a pulpwood. It has also been used for woodenware, millwork and plywood. The Balsam Poplar, *Populus balsamifera*, several other large poplars, and the much smaller Aspens, *Populus tremuloides* and *Populus grandidentata*, have similar uses. Aspen wood is also used in the manufacture of furniture, matches and in cooperage and pulp for making book paper.

Red Gum

Red or Sweet Gum, *Liquidambar styraciflua*, has become of greater importance as a commercial wood by the middle of the 20th Century. This tree ranges from Connecticut to the mountains of Central America, attaining its finest development in the Southeastern United States. The wood is soft and light, but resilient and tough. It is reddish brown in colour, with a fine, straight, close grain. It retains a good polish. It is often stained to imitate cherry, mahogany or walnut. It has been extensively used for veneers, furniture, cabinetwork, interior finish, fancy boxes and cooperage.

The wood is known in Great Britain under the name of Satin Walnut, and the tree is the source of Storax, a medicinal product. The tree has a beautiful conical shape and has been planted widely as an ornamental. The foliage changes to a myriad of colours in autumn that rivals the hard maples. At latitudes of 31-38 deg. North in the southeastern United States the autumn display occurs from October through December and is as startling as that witnessed in the Northeastern portions of the continent.

Sycamore

Platanus occidentalis is a familiar tree throughout the eastern deciduous forest of North America and it has a characteristic bark that peels off in large patches. It is the largest hardwood tree, the wood being tough, strong, hard and very durable. It is light reddish brown with a close uneven grain. A mottled figure with lustrous rays is created when quarter sawed. This is often sold as lacewood. Sycamore has been extensively used for tobacco boxes and other containers because it is odorless. It has also been used for millwork, furniture, interior decoration, butcher blocks, yokes, boxes, crates, woodenware, cooperage, brushes and plywood.

Tulip

Liriodendron tulipifera is one of the largest trees of the eastern deciduous forest of North America, reaching a height of 125-250 ft. and a diametre of 6-14 ft. The wood is known commercially as Yellow Poplar or Whitewood. It is light, soft, and easily worked, with a fine straight grain. It is also stiff and durable although not very strong. It has been used for making boxes, furniture, carriage bodies, musical instruments, woodenware, toys, and boats, light construction and veneers.

Tupelo

There are two species that are known commercially as tupelo, the Black Gum, *Nyssa sylvatica*, and Sour Gum, *Nyssa aquatica*. Black gum is distributed from Michigan to Maine and south to Florida and Texas, while the Sour Gum is restricted to the swamps of the southeastern United States.

The wood is pale yellow with a dense, fine, twisted and interwoven grain. It is soft, tough, light, stiff and resistant to wear. Tupelo has also been used for flooring, tobacco boxes, wheel hubs, woodenware, veneers, railroad ties, handles, pulp, rollers, yokes and piling. The honey manufactured by bees from the blossoms is of the very highest quality.

Walnut & Butternut

Black Walnut, *Juglans nigra*, has been one of the most valuable of North American woods. The tree is large and occurs in the deciduous forest region from Minnesota and Massachusetts to Texas and Florida. It is a moderately hard wood, tough and strong and easily worked. It is very durable.

The colour varies from rich dark brown to purple black. It has a fine even grain and a good figure and is able to take a high polish. Black walnut has been exploited to the point of becoming scarce. It brings a high price, often being sold by the pound. It was being cultivated on farms by the end of the 20th Century.

From the 17th Century onward it has been a principal wood for the manufacture of gunstocks. Other uses included furniture, millwork, cabinetwork, musical instruments, airplane propellers, sewing machines and veneers. In pioneer days it was used locally for barns, fences and light construction. The nut that is produced is very flavourful.

Butternut, *Juglans cinerea*, extends from New Brunswick to Minnesota and south to Arkansas and Georgia. The wood is similar to Black Walnut in nature and uses.

However, the wood lacks the colour and is not as strong. It is used primarily for boxes, furniture, crates, millwork, excelsior and woodenware. Sugar is sometimes obtained from the sap of the butternut and a dye is made from the green husks that contain the fruit.

Misc. Minor Hardwoods

Apple

Pyrus malus wood is usually obtained from old orchards. It is very strong, hard and compact with a uniform close grain. Uses have included the manufacture of tool handles, knobs, pipes, mallet heads, canes, rulers and turnery.

Red Alder

Alnus rubra is the largest of the alders, and is of some commercial importance in Oregon and Washington west of the Cascade Mountains. The wood has a fine even grain, uniform texture and a reddish brown colour. It works and polishes well and gives a good imitation of Black Walnut and Mahogany. Uses include the manufacture of furniture, millwork, handles and novelties.

Blue Beech

Carpinus caroliniana is a small tree of Eastern North America. The wood is heavy, strong and very stiff. No other wood exceeds its suitability for making levers. It is also used for charcoal and tool handles.

Buckeye

Aesculus octandra is a tree of the Middle Western United States that has a soft, light, easily worked wood. Uses have included the manufacture of boxes, excelsior, millwork, piano keys, furniture, trunks and artificial limbs.

Cucumber

Magnolia acuminata is the largest and most numerous of all the magnolias. It ranges in the southern portion of the North American deciduous forest.

Its soft, light, durable wood is used for millwork, boxes, woodenware, excelsior and inexpensive furniture. In the trade it often is sold as yellow poplar.

Dogwood

Cornus florida is a small tree that occurs throughout eastern North America but reaches its optimum development in the southeastern United States.

The wood is very hard and heavy, with a fine, lustrous, close grain. It is used mainly for shuttles for cotton mills and is very resistant to wear. Other uses include wedges, bobbins, mauls, golf-club heads, engraver's blocks and cogs. Only the sapwood has value.

Catalpa

Catalpa speciosa is a small tree native to the lower Ohio Valley. It has been extensively planted in the Middle West.

The wood is very durable and is much used for railroad ties and fence posts. The tree is often planted as an ornamental, and has showy, aromatic blossoms in springtime.

Coffee Tree

Gymnocladus dioica is a tree of the eastern deciduous forest that has a strong and durable wood. It is used for the manufacture of furniture, interior finish, bridges, sills, posts railroad ties and fuel.

Hackberry

Celtis occidentalis is tree of eastern North America. The wood is tough, strong, heavy and moderately hard, and is used mainly in millwork and for boxes, woodenware, furniture and cooperage.

Holly

Ilex opacas is a prominent tree of the Southern coastal area and lower Mississippi and Ohio Valleys of the United States. It is perhaps best known for its leaves and fruit that are associated with the Christmas Season. The tough, close-grained, whitish wood is used for inlays, brushes, fancy articles, and woodenware and is often stained to imitate ebony.

Hornbeam

Ostrya virginiana occurs in Eastern North America and is one of the hardest, toughest and strongest woods known, but it is available only in small quantity. It has been used for carriage parts, levers, handles and fence posts.

Persimmon

Diospyros virginiana is a tree of the southeastern United States. The sapwood is very heavy, tough, hard, strong, elastic and resistant to wear. It has been used mainly for shuttles, boot and shoe findings, golf club heads and other sporting materials and in turnery. The fruit, which ripens in early autumn after a frost, is sweet and delicious.

Sassafras

Sassafras albidum is a small tree of the southeastern and Gulf portions of the United States. The wood is fragrant, durable, soft and light. It has been used for furniture, millwork, cooperage, fence rails and posts and boxes. A root beer is prepared from the roots.

Willow

Salix nigra is the only species of the numerous willows that has attained commercial importance. It is especially abundant in the flood plains of the Mississippi and Ohio Valleys. The smooth wood is soft, light, tough and somewhat strong. It has been used for boxes and crates, plywood cores, excelsior, bats, boats, water wheels and charcoal. The long, slender, pliable young shoots have been used to make wicker baskets and furniture.

Forests of South America

Forests in South America have been decimated by indiscriminate logging and farming expansion, so that by the end of the 20th Century less than 440 percent of the land area still contained forest. The trees are mostly tropical hardwoods (85 percent) with some temperate hardwoods (5 percent) and conifers (6 percent). There are two major kinds of tropical hardwoods.

The most abundant are in the dense humid rain forest that occurs in the Guianas and the Amazon and Orinoco River basins and along the eastern coast of Brazil. This forest is noted for the great number of species and for the size and frequency of the individual trees. There were once over 2,500 different tree species in the Amazon forest alone.

This forest, which is the most extensive body of solid forest in the world, has scant vegetation on the forest floor due to the density of the canopy. Epiphytes and lianas are abundant and characteristic. In drier parts of Argentina and Brazil there is an open deciduous type of tropical forest. Mixed forests of conifers and temperate hardwoods occur along the northern Andes and again in the southern Cordilleras.

Conifer forests are mainly of the Paraná Pine, *Araucaria angustifolia*. These cover large areas in southern Brazil and Argentina, while similar areas, with *Araucaria araucana* the dominant species, occur in Central Chile. In Argentina and Paraguay there are extensive areas of open forests composed mainly of Quebracho, *Schinopsis lorentzii* and *Schinopsis balansae*, important sources of tanwood.

Important Neotropical American Woods

Tropical forests in Mexico, Central America, the West Indies and South America have been harvested for centuries for some of the most high-grade cabinet and furniture woods of commerce. However, these forests have been little exploited as a source of ordinary lumber. The number of valuable species is vast. Brazil alone may have more than 3,000 woody species, including 50 of the most valuable cabinet woods. Some of the more important species are discussed.

Balsa Wood

Ochroma pyramidale is a soft and pithy wood that weighs only 10 lbs per cubic foot. The tree is found in the tropical forests from southern Mexico to northern Peru. Ecuador has furnished most of the world's supply. Balsa is obtained mainly from wild tree, although some plantations have been established. This species grows faster than any other rain forest tree except the Papaya, and reaches a height of 12 ft. in one year. It may be harvested when only two years of age, but usually trees 6-9 years old are selected.

The presence of air in the cell cavities gives this wood its light weight and buoyancy. It has been used for life preservers, buoys, swimming belts, floats, rafts, pontoons, sea sleds, surfboards, toys and motion picture sets and in airplane construction.

The wood also possesses good insulating properties and has been used to line refrigerators, vehicle bodies and the olds of ships. It possesses sound deadening properties and has been used to line ceilings and partitions under heavy machinery to reduce vibrations. The hairs from seeds have been used as stuffing material.

Venezuelan Boxwood

Gossypiospermum praecox has been used as a substitute for European boxwood in the manufacture of engraver's blocks, rulers and other scientific instruments, shuttles, spools, musical instruments, inlays, bobbins and veneers. The supply originates mainly in Venezuela.

Spanish Cedar

Spanish or Cigar Box Cedar, *Cedrela odorata,* ranges throughout tropical America, both as a native and introduced species. It is an important timber tree for local use in tropical America. The wood is reddish brown and aromatic with a straight coarse grain. It has been used for cigar boxes and it is an insect repellent. Other uses include linings for closets, chests, and shingles and as a substitute for mahogany.

Cocus Wood

Brya elbenus is also called American Ebony or Granadillo. The wood is used for knife handles, musical instruments (flutes and clarinets), turnery, inlays and cabinetwork. It is a very hard and durable wood and it polishes well. The supply is mainly from the West Indies.

Cocobolo

Dalbergia *retusa* is a tree that occurs from Mexico to Panama. It is one of the showiest and most strikingly coloured of the exotic woods. The heartwood *is* orange-to-orange red in colour, streaked with jet black. It is very hard, tough and strong, and has been used for instruments, knife and umbrella handles, steering wheels, inlays, lacquer and turnery.

Crabwood

Carapa guianensis attains its best form in the Guianas, although it occurs also in other parts of South America and the West Indies. This very strong and hard, brown wood is used as a substitute for mahogany.

Greenheart

Ocotea rodioei is native to Guyana. The greenish-brown wood is used for structural timbers as it is one of the strongest and is hard, heavy, tough, resistant to decay and insect injury and elastic. It has been widely used in Europe for bridges, piles, wharves, paving blocks and shipbuilding. It is also used for shafts, spokes and fishing rods.

Lancewood

Two trees of the West Indies constitute the lancewood of commerce: Lancewood, *Oxandra lanceolata* and Degame or Lemonwood, *Calycophyllum candidissimum.* These give a yellowish wood with a fine close grain and which is very tough, strong and elastic. They are used for fish poles, spars, shafts, whips, ramrods, bows, cues, lances and turnery.

Letterwood

Piratinera guianensis of Guyana yields a very heavy ornamental wood. The colour is brown with peculiar black markings. It has a close, fine and lustrous grain. It has been used for canes, umbrellas, violin bows, drumsticks, veneers and inlays.

Lignum Vitae

Two species, *Guaniacum officinale* and *G. sanctum* supply this important wood. The trees occur in southern Mexico, Central America and the West Indies. This is one of the hardest of commercial woods, and it is naturally tough, strong and resistant. The colour is dark brown, streaked with black, with a very fine, intricately woven grain. It also contains a resin that acts as a natural lubricant and preservative. It has been used for bearings or bushing blocks for steamship propeller shafts. It is also used to manufacture bowling balls, pulley blocks and conveyors, instruments and furniture. It was once believed that Lignum Vitae possessed remedial powers for many human diseases, and that is how it got its name "lignum vitae" or "wood of life." A gum resin, Guaiacum, is used in medicine and occurs in the form of tears excreted from the living tree.

West Indian Locust

Hymenaea courbaril is a very important timber tree in tropical America. The wood is hard and tough, and it is used for general carpentry, furniture shipbuilding and cabinetwork. This species supplies South American Copal.

Mahogany

Several species of mahogany make up the most important export wood in tropical America, and they are the source of the world's most valuable timber and premier cabinet wood. These are *Swietenia mahogany*, the West Indian Mahogany or Spanish Mahogany was used for woodwork by 1514 A.D. This species that ranges from the Florida Keys to the West Indies, was the first to be harvested commercially and the first to be exploited. It has been introduced into Central America and other tropical countries. Another species, *Swietensia macrophylla*, occurs from Yucatan to northeastern Colombia and Venezuela and along most of the southern effluents of the Amazon in Brazil, Peru, and Bolivia. Although there are a few other species, for practical purposes all the native mahogany cut in North and South America probably belongs to this species.

The cathedral of Santo Domingo that was built in 1550 contains some magnificent mahogany carvings that are still in top condition. The early Spanish explorers utilized the timber for shipbuilding, and it was also used for building construction in England by 1680. From the beginning of the 18th Century it began to be used as furniture wood. During the years 1750 to 1825 great craftsmen of furniture, Chippendale, Adam, Hepplewhite and Sheraton, used mahogany.

Mahoganys are ornamental evergreen trees that reach a height of 40-50 ft. with large buttresses at the base. The trees occur scattered through rich moist forests about one to the acre. The wood is reddish brown with a crooked grain. It is very heavy, strong, hard enough to resist indentations, but easy to work with and it polishes and glues well. It has been used for furniture, fixtures, musical instruments, millwork, cars, ships and boats, caskets, airplanes, foundry patterns, veneers and plywood. There are many substitutes and imitations, but the supply of true mahogany is still abundant, especially as plantations production is proving successful.

Majagua

Also known as Blue Mahoe, *Hibiscus elatus* of the West Indies is a hard and slightly aromatic wood with a lustrous, richly variegated, open grain. It has been used for cabinetwork, carriages, gunstocks, fishing rods and ship's knees. It is the source of Cuba Bast.

Mora

Mora excelsa of Guyana, Venezuela and Trinidad yields a brown wood that is very hard, tough and even more durable than teak. It has been used primarily in Europe for shipbuilding, platforms, railroad ties and all types of heavy construction.

Prima Vera

Sometimes called White Mahogany, *Cybistax donnell-smithii* occurs in southern Mexico, northern Honduras, El Salvador and Guatemala. The cream coloured wood is used for furniture and fixtures, millwork, ships. Boats, cars and veneers. The tree is also very ornamental with many yellow flowers that appear before the leaves.

Purpleheart

Peltogyne paniculata occurs in Brazil, the Guyanas and Trinidad. The brown wood is very hard, tough, strong and durable. It gradually turns purple on exposure to the air. It has been used for heavy construction, furniture, billiard tables, cues, fishing rods, inlays and turnery.

Rosewood

Brazilian Rosewood, *Dalbergia nigra,* is the best known of several different species. The dark purple, almost black, wood is often striped and has a coarse, dense, even grain. This is one of the finest cabinet woods and also finds use for scientific instruments, furniture, cars, pianos, sporting and athletic goods, brushes and handles.

West Indian Satinwood

West Indian Satinwood, *Zanthoxylum flavum,* has been used to make furniture for a long time. It was especially valued in Great Britain by such builders as Sheraton, Adam and Hepplewhite. The creamy or golden yellow wood is smooth and lustrous and slightly oily. It has a very close, dense and even grain. Other uses have included millwork, musical instruments, caskets, brushes, inlays, cabinetwork and veneers.

European Forests

The forests in Europe have occupied about 30 percent of the land area for quite some time as they are rigidly managed. About 70 percent of the forests are coniferous, 24 percent are temperate hardwoods and less than 2 percent are mixed forests. The coniferous woodlands are especially numerous in the northern portion of the continent, while the hardwood and mixed forests are found in Southern and Western Europe. The original forest cover of Europe has been greatly depleted from long utilization and the necessity of clearing land for agriculture and industry. In the British Isles only about 5 percent of the original forest was left by the end of the 20th Century. In France, Spain, Belgium, Italy and Greece about 10-20 percent was left. Over half the forests in Sweden and Finland are original, and together with Russia, are the most heavily forested regions in Europe.

European Conifers

There are many genera of American origin found in European forests. The principal conifers are the Scotch Pine, *Pinus sylvestris* and Norway Spruce, *Picea abies.* These furnish the woods referred to as Yellow Deal and White Deal, respectively. Other conifers include the Cluster Pine, *Pinus pinaster,* the Stone Pine, *Pinus pinea,* the Silver Fir, *Abies alba,* Larch, *Larix deciduas, and* Yew, *Taxus baccata. American white pine and Douglas fir are also extensively planted.*

European Hardwoods

Oak trees represent the most important hardwoods in Europe. The principal species are *Quercus cerris*, *Quercus robur* and *Quercus petraea*. Other hardwoods include the Black Alder, *Alnus glutinosa*, European Ash, *Fraxinus excelsior*, European Beech, *Fagus sylvatica*, European Birches, *Betula pendula* and *Betula pubescens*, European Cherry, *Prunus cerasus*, European Chestnut, *Castanea sativa*, European Elm, *Ulmus procera*, Hazel, *Corylus avellana*, European Holly, *Ilex aquifolium*, European Hornbeam, *Carpinus betulus*, European Lime, *Tilia cordata*, European Maple, *Acer* pseudoplatanus, European Plane, *Platanus* orientalis, Rowan, *Sorbus aucuparia*, Blackthorn or Sloe, *Prunus spinosa*, and European Willow, *Salix alba*.

The main European trees that have been imported into North America include the Boxwood, Walnut, Briar and Olive.

Turkish Boxwood, *Buxus sempervirens*, of Southern Europe, Asia Minor and Northern Africa, has practically disappeared due to over exploitation. The wood of this tree is very hard, with a fine, dense, uniform grain and a smooth, lustrous texture. It was sued for blocks, engraving, rulers and other instruments, shuttles, bobbins, firearms, whips, canes, inlays, umbrellas and turnery.

English Walnut, *Juglans regia*, also called Circassian Walnut, occurs naturally from the Black Sea region across Asia Minor and Persia to northern India. The wood is hard and beautifully figured and it takes a high polish. Wood obtained from the burls is particularly valuable. It has been used for fixtures and furniture, millwork, musical instruments, firearms, cabinetwork, turnery and veneers.

European Olive, *Olea europaea*, has wood that is suitable for brushes, canes and turnery, and of course the fruit is invaluable for its oil. Briar Root, *Erica arborea*, has been traditionally used for tobacco pipes.

Asian Forests

By the end of the 20th Century forests covered less than 19 percent of the land area in Asia. These forests are most abundant in the northern and eastern parts of the continent and they are very sparse in the southwest. Much of the original forest area has been exploited to complete destruction. This is particularly evident in China, where centuries of cultivation have not only destroyed the forests, but much of the arable land as well because of over cultivation and subsequent erosion.

Asian Conifers

Conifers make up about 40 percent of the forest area. These are characteristic of most of Siberia and also occur in the Himalayas and the mountains of Asia Minor, Japan and China. European species are found in the western part of the continent, giving way to more distinctive Asiatic species of the same genera in the east. Pine, Spruce, Fir, Juniper, Cedar, Larch and Yew are the principal groups.

Asian Hardwoods

Temperate hardwoods make up about 26 percent of the forests. These and mixed forests are found in the southern portions of Russia, Afghanistan, Iran, Asia Minor, China and Japan. European species are found in the western part of the continent, giving way to more distinctive Asiatic species of the same genera in the east. Maple, Ash, Basswood, Poplar, Alder, Birch, Walnut, and Oak are the principal groups. Tropical hardwoods make up about 30 percent of the total forest area and are found south of the Himalayas. In many areas they comprise all of the woody species.

This is particularly true in Sri Lanka, Thailand, the East Indies and Malaya. As in other tropical forests the number of species is very large. India is estimated to have about 2,000 and Japan 1,000 different species. Many of the Asiatic forests are dominated by teak, while others are composed mainly of members of the Dipterocarpaceae and Leguminosae.

The former family comprises over 75 percent of the tree species in the Philippines. A large number of these Asiatic woods enter into world commerce, the most important of which include Ebony, Padouk, Satinwood and Teak..

Ebony

There are several species of tree that furnish a wood known as ebony, the most important of which is the Macassar Ebony, *Diospyros ebenum*, found from India to the East Indies. The wood is black with brown stripes. It is very hard and heavy and has a fine grain. It takes a high polish and has been used since ancient times as a cabinet wood. Other uses include canes, whips, umbrellas, piano keys, sporting and athletic goods, inlays, handles, turnery and veneers.

Padouk

Also called Burmese Rosewood, Pterocarpus indicus is a very ornamental tree. The lustrous wood is red in colour with black stripes.

It is hard and durable and polishes well. Padouk has been used for furniture, fine cabinetwork, car construction, millwork, veneers and turnery. The tree has been introduced into Honduras.

Asian Satinwood

East Indian Satinwood, *Chloroxylon swietenia,* has a hard yellowish or dark brown heartwood that is sometimes mottled. It has a satin like luster and a fine, dense, even grain. Satinwood is used for cabinetwork, furniture, brushes and veneers.

Teak

Tectona grandis is native to Southeastern Asia and Malaya. It is one of the most durable of woods and is a very important timber of the tropics. The wood is hard and it does not warp, split, or crack. Thus it is valuable for general construction. It is also very resistant to termites and decay. The colour is yellowish brown and greasy to the touch. It has been widely used in ship and boat building, and for cabinetwork, furniture, millwork, piles, railway cars, greenhouses and flooring. The tree has been introduced into the West Indies and Panama where it has shown phenomenal growth. One tree attained a height of 64 ft. and a diametre of 18 in. in less than 16 years.

Other important trees of tropical Asia that supply valuable wood include the Acle or Pyinkado, *Xylia xylocarpa,* Deodar, *Cedrus deodara,* Sal, *Shorea robusta,* and Sissoo, *Dalbergia sissoo.* These have all been used for general construction Additional species used for cabinet manufacture include Moulmein Cedar, *Cedrela toona,* Laurelwood, *Calophyllum inophyllum,* Rosewood, *Dalbergia latifolia* and Asian Sandalwood, *Santalum album.*

African Forests

The African Continent consists of primarily of deserts or semi-deserts, so that forests cover only about 10 percent of the area. These begin south of the Sahara Desert. Tropical hardwoods are in the majority with over 95 percent of the forests. Temperate hardwoods account for less than 4 percent and conifers less than one percent, the latter being restricted to lands along the Mediterranean and the higher mountains in Central, Eastern and Southern Africa.

Two main types of tropical forests occur. First, the extensive dense rain forest that covers much of equatorial Africa, especially the West African coast and the Congo Basin. This region has a rainfall of 60-160 in., which is uniformly distributed throughout the year. Salient characteristics of this forest are the mangrove swamps along the coast. An open park like forest exists in regions where rainfall is only 30-40 in. Large areas of this type of forest occur in the northeastern and southern parts of Africa, especially in Angola and Zimbabwe.

There are many important kinds of woods that resemble those of tropical America and Asia, but they have not been as extensively exploited. African Mahogany, *Khaya senegalensis*, a West Coast species, is exported in some quantity. This substitute for true mahogany has a wood that is beautifully figured and coloured. The coastlands from Senegal in the west down through Mozambique in the east contain scattered stands of palm thickets prominent among which is the Senegal Date, *Phoenix reclinata*.

An era of forest exploitation began in the equatorial rainforest at the end of the 20th Century, which is threatening the existence of many native species. Efforts to curb this devastation are met with resistance from the local populations that are facing continuous political unrest.

Australian & Oceania Forests

In Australia only about 6 percent of the land area is covered in forests. In New Zealand about 26 percent is forested and in Oceania 70 percent. The most extensively forested regions are New Guinea with 80 percent, Samoa with 70 percent and Tasmania with 64 percent. Conifers comprise about 4 percent of the Australian forests and 11 percent are temperate hardwoods with the rest being tropical hardwoods. Conifers are found in New South Wales, Queensland and Tasmania and temperate hardwoods only in Tasmania. Tropical hardwoods occur in all the states except Tasmania.

These forests are dense and resemble the size and density of those along the Pacific Coast in North America. However, the species differ from those found in other parts of the world. Most trees are either eucalyptus or acacias. Over 70 commercial species of Eucalyptus exist, the most important of which are the Karri, *Eucalyptus diversicolour*, and Jarrah, *Eucalyptus marginata* of Western Australia. Karri may reach a height of 300 ft. with a clear length of 180 ft. Other valuable trees are the Blackwood, Acacia melanoxylon, and Silk Oak, *Grevillea robusta*. The predominant coniferous species is Hoop Pine, *Araucaria cunninghamii*.

About 68 percent of the forests in New Zealand are coniferous and the rest hardwoods. The main species is the Kauri Pine, *Agathis australis*, one of the largest timber trees in the world. The wood is strong and durable and generally free from knots. It has been exported in large quantities. An important resin is also extracted from the kauri. Other valuable species include the New Zealand White Pine, *Podocarpus dacrydioides*, Totara, *Podocarpus totara*, New Zealand Red Pine, *Dacrydium cupressinum* and several species of beech.

Forests in Oceania are composed entirely of tropical hardwoods, with few being of importance for export. On the small island of Norfolk off the east coast of Australia, the Norfolk Pine, *Araucaria heterophylla* (*excelsa*), was being harvested in considerable quantity in the 20th Century for timber. Its rapid growth enabled a viable industry on the island, only five miles in diametre.

Gums & Resins

Gums

True gums are formed from the disintegration of internal plant tissues, mostly from the decomposition of cellulose in a process called gummosis. Gums contain high amounts of sugar and are closely allied to the pectins. They are colloidal and soluble in water, either dissolving entirely or swelling, but they are insoluble in alcohol and ether.

They exude naturally from the stems or in response to wounding of the plant. Commercial gums arrive in the market in the form of dried exudations. Gums are especially common in plants of dry regions. They are used primarily as adhesives, and are also used in printing and finishing textiles, as a sizing for paper, in the paint and candy industries and as drugs. Three important commercial plant gums are gum arabic, gum tragacanth and karaya gum.

Gum Arabic

A dried gummy exudate obtained from *Acacia senegal* and related acacias. These are small native trees of arid northern Africa and are extensively cultivated in the Sudan. The trees are tapped between February and May when the fruits are ripe. Transverse incisions are made with a small ax and thin strips of the outer bark are torn off. The gum slowly exudes as a viscous liquid, collects in a drop and hardens. After 3-8 weeks these "tears" are collected. The sun bleaches them, and the impurities are removed before marketing.

The Egyptians used gum arabic as early as 2,000 B.C. Sudan gum has been an article of commerce since 100 A.D. Several kinds reach world markets. Kordofan or Hashab gum has been exported from the region around Cairo and Port Sudan, while Senegal gum originates north of the Senegal River. Gum arabic is slowly and completely soluble in cold water and has a high degree of adhesiveness and viscosity. Most has been used in textile, mucilage, paste, polish and confectionery industries and as a glaze in painting. In medicine it has been used as an emulsifying agent and as a demulcent.

Gum Tragacanth

This is formed by a transformation of the pith and medullary-ray cells into a mucilaginous substance that exudes naturally or after the bark has been punctured or excised. It comes from *Astrangalus gummifer* and other species of that genus. They are thorny shrubs of the arid regions of Western Asia and Southeaster Europe. The gum is allowed to dry on the bark before it is collected. It reaches the market in one of three forms: tears, which are the dried natural exudate; vermiform gum, which consists of narrow twisted coils or strings; and flakes, which are ribbonlike pieces. Most of the commercial supply originates in Iran and Turkey. Gum tragacanth has been used in calico printing and for other industrial purposes. It is one of the oldest drugs and was known since 300 B.C. In modern medicine it has served as an adhesive agent for pills and troches and for the suspension of insoluble powders.

Karaya Gum

Karaya gum has been used as a substitute for gum tragacanth, and several million pounds were imported annually from India by the mid 1900's. It is used in the textile, cosmetic, cigar, past and ice cream industries. It is obtained from *Sterculia urens,* a large tree in central India. Incisions are made into the heartwood and the gum oozes into these and accumulates as large irregular knobs. They are then collected, sorted and graded. The gum enters the trade in the

Other Gums

There are many other plants, which produce gums of some commercial importance. Gum ghatti, obtained from *Anogeissus latifolia,* a large tree native to India and Ceylon, is used as a substitute for gum arabic. The leaves are used for tanning. *Feronia limonia* and *Cochlospermum religiosum* in India, Burma and Java yield gums that are also used in place of gum arabic. The Asiatic *Cycas ciccinalis* is the source of cycas gum. The carob, *Ceratonia siliqua,* produces tragasol, a mucilaginous hemi cellulose occurring in the pods. Mesquite gum in North America is obtained fro *Posopis juliflora, P. glandulosa* and other species, while cherry gum is found in various species of *Prunus.*

Resins

Resins are formed as oxidation products of various essential oils and are very complex and varied in chemical composition. The resin is usually secreted in definite cavities or passages. It frequently oozes out through the bark and hardens on exposure to air. Tapping is usually necessary in order to obtain a sufficient amount to be of commercial value. Commercial resins are also frequently collected from fossil material. Resinous substances may occur alone or in combination with essential oils or gums. Resins, unlike gums, are insoluble in water, but they dissolve in ether, alcohol and other solvents. Resin production is widespread in nature, but only a few families are of commercial importance. These include the Anacardiaceae, Burseraceae, Dipterocarpaceae, Guttiferae, Hammamelidaceae, Leguminosae, Liliaceae, Pinaceae, Styracaceae and Umbelliferae. The exact botanical origin of a resin is often hard to trace, especially in the case of fossil and semi fossil types.

Resins probably serve the plant by preventing decay, which is due to their high antiseptic qualities. They may also lower the amount of water lost from the plant tissues.

Resins have certain characteristics that render them important to industry. Their ability to harden gradually, as the oil that they contain evaporates, makes possible commercial varnishes. The resins are dissolved in solvents and surfaces are painted with the mixture. As the solvents and oils evaporate, a thin waterproof layer of resin remains. Resinous substances have been utilized for waterproof coatings, and also for decorative coatings for millennia. Ancient Egyptians varnished their mummy cases and the utilization of lacquer in the arts has been practiced in China and Japan for centuries. The Greeks and Romans were familiar with many of the same resinous materials that are used today (e.g., mastic, amber, sandarac). Another property of resins that is of industrial importance is their ability to dissolve in alkalis to form soap. Resins are also used in medicine; for sizing paper; as a stiffening material for mats; in the preparation of sealing wax, incense and perfumes; and for many other purposes as well.

It is difficult to classify resins because the same term is often used for very diverse materials. In commerce, resins are often referred to as gums, while such terms as varnish resins, hard resins, spirit varnishes, balsams, gum resins, damar resins, soft resins, and many others are used quite indiscriminately. The chemical differences between the various groups are much more definite. In this section three groups that will be distinguished are Hard Resins, Oleoresins and Gum Resins.

Hard Resins

These contain only a little if any essential oil. They are usually solid, quite transparent, brittle substances with no particular odor or taste. They are readily fusible and burn in air with a smoky flame. They are nonvolatile and very poor conductors of electricity. They become negatively charged with friction. The hard resins constitute the best source of varnishes, due to their low oil content and the readiness with which they dissolve in alcohol. The most important commercial resins, such as the copals and damars, belong to this group. Hard resins are also used in paints, inks, plastics, sizing, adhesives, fireworks and other products.

Copals

Copals comprise a considerable group of resins of recent, semi fossil and fossil origin that are found in many tropical and subtropical areas of the world. The word "copal" is of Mexican origin. Many of the harder copals are called animes, especially in northern Europe. The copals insignificant quantities of oil and they yield a hard elastic varnish, which has been widely used for outdoor work. Several types are know which are diverse in character and source

East African Copals

Zanzibar copal and the closely related Madagascar and Mozambique copals are derived from *Trachylobium verrucosum*. Zanzibar copal is the hardest of all resins except amber and is very valuable. The resin exudes naturally from the trunk, branches and fruit. However, most of the commercial supply was obtained from semi fossil material derived from still living trees, and fossil material from trees that no longer exist. Fossil copal when removed from the ground is covered with a crust of oxidized material. After this has been removed, the copal shows a typical surface known as "goose skin," which consists of large and small excrescences. The interior is clear and transparent and varies from yellow to brownish red in colour.

Inhambane Copal is obtained from *Copaifera conjugata*, a valuable timber tree in coastal Southeast Africa; but it is of little commercial value.

West African Copals

There are a large number of hard resins among the West African copals. They are usually designated by the name of the region from which they were obtained. The most important of these are the Congo, Angola, Sierra Leone, Acere and Benin copals.

Congo copal is the most important because it is very hard and has been extensively exploited in the 20th Century. It is derived from *Copaifera demeusii* and *C. mapane*, which are characteristic trees of the Congo Basin. Although living trees have furnished some of the supply, the greater part was obtained from the ground or from watercourses, and was more or less fossil in nature. The white and red Angola copals are derived from the same two species.

Sierra Leone copal is a light-yellow, hard and brittle resin that is obtained from *Copaifera copallifera* and *C. salikounda*. Live trees are wounded and the resin exudes and hardens in the form of globular tears. Some fossil material is also obtained. Because of the value of this copal, the Sierra Leone government has protected the trees.

Accra and Benin copals are probably derived from *Daniella ogea* and related species. These large trees are found in the coastal forests of Liberia, the Gold Coast and Nigeria. The resinous exudation, known locally as ogea gum, has gained prominence as a varnish resin.

Kauri Copal

This copal is one of the most valuable of the hard resins. It is obtained from the kauri pine, *Agathis australis*, which is the largest and most important tree in New Zealand. The copal is chiefly fossil in nature and is dug up on ridges and in swamps and bogs. "Swamp gum" furnishes the bulk of the supply, and ranges in size from pieces that are 1-2 inches in diametre to lumps weighing 100 lbs. "Range gum" yields the best grade of kauri. Tapping living trees gets an inferior "Bush Gum". Kauri is yellow, transparent and very hard. It is an exceedingly valuable varnish resin, especially for marine and outside work. It was also used in making linoleum. Kauri constituted one of the chief exports of New Zealand in the mid 20th Century.

Manila Copal

The first shipments of this important copal were made from Manila and the name persisted, although by mid 20th Century 75 percent of the product was shipped from Indonesia. The source of all the East Indian, Philippine and Malayan copals, of which there are many different kinds, is *Agathis alba*. This is a tall conifer reaching a height of 200 ft.

The resin exudes naturally, and is also obtained by systematic tapping. Some of the supply was derived from fossil material and consisted of large, irregular, angular, milky pieces with a yellowish interior. Hard, semi hard and soft copals are included among the many different kinds in the trade. Pontianak copal, a semi fossil type from Borneo, is the hardest variety and was especially popular in the United States. Manila copal varnishes are durable, but they do not adhere strongly to surfaces and they are not very brilliant. They are used chiefly for interior work and enamels. These copals are often erroneously called damars, although they are quite distinct from the true damars.

South American Copals

The tall locust tree in South America, *Hymenaea courbaril*, which is found in Brazil and other portions of Neotropical America, is the main source of this copal. The stems, twigs and even the fruits exude a large amount of resin, which trickles to the ground. The commercial resin is collected from the base of living trees, and former trees as well, and is marked as Demerara or Para Copal. It is the softest of all copals and therefore the least valuable.

Damars

There is confusion in the term "damar." The word is of Malayan origin and is used by the natives to indicate a torch made of decayed wood and bark, mixed with oil and powdered resin, wrapped in leaves and bound with strips of rattan. At one time it did not refers to any specific tree or resin. However, the word came to be a collective term for a great variety of hard resins of different origin, and included even kauri and manila copal. In the commercial trade the term "damar" restricted to resins that re obtained from members of the Dipterocarpaceae. A few resins from species of the Burseraceae are also included. This distinction should be maintained for the true damars are very different chemically from the various coniferous resins. Unlike Manila copal, damar is insoluble in chloral hydrate, but completely soluble in alcohol and turpentine.

The trees that yield damars are characteristic of all Southeastern Asia, and are particularly abundant in Malaya and Sumatra. Although all members of the Dipterocarpaceae secrete resin, only a few species are of commercial importance. These are found mainly in the genera *Balanocarpus*, *Hopea* and *Shorea*. Damars are especially important in Malaya and are obtained by tapping the trees.

The most important Malayan varieties are Damar Mata Kuching from *Hopea micrantha* and related species. Damar Penak from

Balanocarpus heimii, and Damar Temak from *Shorea hypochra*. The principal damars of India are saldamar from *Shorea robusta*, white damar from *Vateria indica*, and black damar from *Canarium strictum*. Damars also come from Borneo, Java, Sumatra, Thailand and Cochin China. Batavian damar, a product of *Shorea wiesneri*, is the best commercial variety. The so-called East India resins are products of the same trees that yield damars. They are older and harder and are often gathered from the ground or from watercourses.

Damars were used principally in spirit varnishes and the manufacture of nitrocellulose lacquers. Damar varnishes adhere better than Manila copal varnishes, but are softer and less durable. They are particularly well adapted for varnishing paper because of their luster and light colour. They have also been used for indoor work and in histology.

Amber

Amber is a fossil resin found mainly along the shore of the Baltic Sea and to a lesser extent in Mexico, Central America and Caribbean Islands. The principal source of Baltic amber was the now extinct pine, *Pinus succinifera* that flourished on the shores of a former sea in the Eocene. Amber is very hard and brittle. It occurs in several forms, the most important of which is succinite. Some of these forms are transparent and others are almost opaque. The colour varies from yellow to brown and even black. Amber assumes a high polish and becomes negatively charged when rubbed. It also gives off a characteristic aromatic odor. This substance has been known for thousands of years. The Swiss Lake Dwellers were familiar with it and it was highly prized by the Greeks and Romans. It has always been used for beads and other ornamental purposes and is often carved.

The chief use of amber is for the mouthpieces of pipes and cigarette and cigar holders. The darker grades yield a valuable varnish, but this is too costly for general use. Amber is also used to increase the elasticity of rayon fibres and as the source of an essential oil. It is of scientific interest that there are often found imbedded in amber the remains of plants, insects and other objects that existed at the time the fresh resin was exuded from the pine trees.

Lacquer

Lacquer is a natural varnish that exudes from various trees in Asia, and great quantities of it are used in the Orient for ornamental purposes. The principal source is the lacquer tree, *Rhus verniciflua*, native to China but cultivated in Japan. The trees are systematically tapped. The exudate is a milky liquid, which darkens and thickens rapidly on exposure to air. However, it can remain unchanged for long periods by storage in closed containers. It is filtered before use. When applied as a varnish, the thin film rapidly hardens in a moist atmosphere, in part due to oxidation. Lacquer imparts a remarkable protection as it is unchanged by acids, alkalis, and alcohol or heat up to 160 deg. Fahrenheit. Pigments may be mixed with the lacquer before drying.

The lacquering process originated in China many centuries before Christ, and reached its highest development during the Ming dynasty (A.D. 1368-1644). In Japan the first records are from the 4th Century, when lacquer was used for many purposes. The earliest examples known belong to the 6th Century. The art reached its height during the 17th Century, although much fine work was produced as late as the 19th Century. In Japan the quality of lacquered goods is of the highest excellence. Gold has often been used as a colouring. The process of lacquering is very complicated and tedious. Sometimes 300-400 coats are applied and the whole operation requires several years to complete. This technique was kept secret for many years.

In Burma lacquer is obtained from *Melanorrhoea usitata*. It dries more slowly than Japanese lacquer, but has been used in an attempt to build an industry in Burma. Natural lacquers are also obtained in Taiwan and Vietnam, with *Rhus succedanea* being the source in the latter.

Shellac

True shellac is not a direct plant product but deserves discussion here. It is made from stick-lac, a resinous substance secreted on the twigs of many trees by an insect, *Tachardia lacca*. The lack insect derives its food from the sap of the trees and secretes the resin as a kind of cocoon for its own protection and that of its offspring. Although there are about 40 species that may serve as hosts for the insect, only seven are of importance. These are often cultivated. They include *Butea monosperma, Schleichera oleosa, Zizyphus xylopyrus, Ficus religiosa, Acacia nilotica, Cajanus cajan, and Zizyphus jujuba. Butea monosperma* was used as a host as early as 250 A.D. At first a valuable red dye, which was obtained from the insect, was the only sought after product. However, since 1590 A.D. the resinous excretions have been more important. Native workers prepare most of the shellac of industry. India furnished over 97 percent of the total output by the mid 1950's, the rest being from Burma, Thailand and Vietnam. The crude stick-lac is removed from the twigs and soaked in water to extract the red dye. It is then dried and powdered to a granular consistency. This seed-lac is melted and thin sheets of it are hardened. These are broken into the semitransparent, brittle, orange-red flakes which make-up shellac. If the melted seed-lac is poured out in droplets, it hardens into the thick round pieces known as button-lac. Shellac is often bleached.

At one time shellac had many industrial uses, but synthetic substitutes made from petroleum have greatly competed with it. True shellac can be molded readily and was the most satisfactory material for the manufacture of phonograph records. It is an insulator of the highest grade and was once extensively used in the electrical industry.

It was the principal spirit-varnish resin yielding a tough film with a smooth finish, which is also capable of a high polish. Shellac varnishes cannot be used out-of-doors because they are not water resistant. Shellac was also used in making sealing wax, drawing inks, some watercolours and nitrocellulose lacquers; for sizing papers; for stiffening felt hats; and, in India, for many ornamental purposes.

Acaroid Resins

These are grass-tree resins obtained from Australian trees belonging to the genus *Xanthorrhoea.* These plants are among the few monocotyledons, other than palms, which have an arboreal habit. They consist of a short woody stem, composed of the old leaf bases, surmounted by a tuft of long rush like leaves. The resin collects around the bases of the old leaves and is removed by beating the stem.

The yellow acaroid resins reach the market in the form of elongated or round reddish-brown pieces. They are obtained chiefly from Xanthorrhoea hastilis. Red acaroid resins from *X. tateana*, *X. australis*, and related species are much more common. They consist of uneven pieces of a brownish colour. The acaroid resins are used in the manufacture of sealing was, gold, size, and spirit varnishes for use in coating metals; as a substitute for rosin in paper sizing and inks; as a mahogany stain; as a source of picric acid; and in medicine.

Sandarac

This is a soft pale-yellow resin obtained mainly from Tetraclinis *articulata* (= *Callistris quadrivalis*), a small tree in Northern Africa. Australian species of *Callitris* are also a source of sandarac. The resin is formed between the inner and outer layers of the bark, and is excreted inn the form of small tears, which quickly become opaque. Sandarac yields a hard, white, rather brittle spirit varnish that is especially useful for coating labels, negatives, cardboard, leather and metals. It was once used in medicine and was well known to previous civilizations.

Mastic

This is a very old resin that was known at least as far back as 400 B.C. The most useful variety is Chios mastic, derived from *Pistacia lenticus*, a small tree from the Mediterranean area. Although the resin exudes naturally, removing strips of bark enhances the flow. Some of the resin adheres to the trunk in the form of long, ovoid, pale-yellow, brittle tears, while the remainder falls to the ground. Bombay mastic, which consists of large irregular pieces of a dull, milky colour, is from *P. cabulica*. Mastic yields a pale varnish used for coating metals and pictures, both oils and watercolours. It is also used in lithographic work, in perfumery, in medicine, and as cement for dental work. It is one of the most expensive and high-grade resins.

Dragon's Blood

Included here are various deep red materials of a resinous nature. Sumatra dragon's blood is from *Daemonorops draco*, a climbing rattan palm of Eastern Asia. The dark reddish-brown resin forms as small granules on scaly fruits. It is used mainly in the manufacture of red spirit varnishes for metals and in making zinc line engravings.

During the 18th Century Italian violinmakers used dragon's blood in their varnishes. Socotra dragon's blood is a resin that exudes from the stem of *Dracaena cinnabari* of Western Asia. It has been also used for varnishes, dyes and stains. Dragon's blood is sometimes secured from tropical American species of Dracaena and other Asiatic species of *Daemonorops*.

Kinos

Gum kinos are produced by several plant species. Malabar kino is derived from the dried juice of *Pterocarpus marsupium*, a large tree in India. Trees are tapped and the juice is boiled down. It reaches the market in the form of small, brownish-red, brittle pieces. West African kino is a red resin from *P. erinaceus*. Bengal kino is from *Butea monosperma*.

Some Australian species of *Eucalyptus* are also sources of gum kino, but mainly *Eucalyptus camaldulensis*. The kino is secreted in cavities between the wood and bark, and oozes out after incisions have been made. In the air the resin hardens into a solid reddish mass.

Several tropical American trees, mainly Dipteryx odorata and Cocoloba uvifera, also yield kinos. Kinos have found their main use in medicine for throat disorders, and are sometimes used in tanning.

Oleoresins

A considerable amount of essential oils are contained in oleoresins in addition to the resinous materials. Thus, they are often liquid in nature. They have a distinct aroma and flavour. Among the oleoresins we find the turpentines, balsams and elemis. There is only a very slight distinction between these groups and often a confusion of names.

Turpentines

These are oleoresins that are obtained almost exclusively from coniferous trees. They are viscous, honey like liquids or soft and brittle solids. The resin is secreted and stored in ducts near the cambium layer and exudes naturally as a soft, sticky substance, often called pitch. For commercial use crude turpentine is obtained by tapping the trees. On distillation turpentines yield the essential oil or spirits of turpentine, and rosin, both of which are useful products around which an important industry had been built in the 20th Century. The turpentine or navel-stores industry was one of the oldest of the forest industries. The Trojans and Greeks were familiar with pitch and its uses, and it is mentioned in the Bible. The industry was valued at 35-40 million US dollars annually by the mid 20th Century. The United States led in production with over half of the total output. Turpentine and rosin are also produced in Europe and India and Vietnam.

Turpentines of Minor Importance

Turpentines from various species of conifers have been frequently used in their natural state for purposes other than the production of naval stores. Some of them have been important in industry, in the arts and particularly in medicine for over five centuries. Among the more common are the following:

Canada Balsam

This is actually turpentine rather than a balsam, as the name seems to indicate. It is secreted by the balsam fir, *Abies balsamea,* of northeastern North America. The resin collects in elongated blisters on the bark and only small amounts are obtainable. It was estimated that a tree could yield from 8-10 oz. per year. Collectors use a pot with a spout cut at an angle. This is forced into the blisters and held in place while the balsam drains out. The balsam is a viscid yellowish or greenish substance. It main use has been as a mounting medium for microscopic work and as cement for optical lenses. It is very transparent and has a high refractive index, which results in a minimum of dispersed light. The medicinal value of Canada balsam was known as early as 1607. It was used as an irritant, stimulant and antiseptic, and is a component of collodion and some plasters. It has also been used as a fixative for soap and perfumes. Oregon Balsam is obtained from the Douglas fir and it has similar properties and uses.

Spruce Gum

This is the natural exudation of various species of spruce of North America, with *Picea rubens* being the main source. It is usually the result of an injury to the sapwood. The thin, clear, bitter, sticky oleoresin is secreted in blister like cavities in the bark or in longitudinal fissures in the wood. It hardens on exposure to the air and is collected when hard or semi soft. Before the advent of chewing gum about 500,000 lb. were used each year as a substitute. The gum softens in the mouth and assumes a reddish colour. It has a pleasing resinous taste.

Venetian Turpentine

This turpentine is obtained from the European larch, *Larix decidua*, a widespread tree in the mountains of Central Europe. Unlike other conifers, the resin ducts are located in the heart of the tree so that holes are bored in order to obtain the resin. The trees are tapped in the spring. Venetian turpentine has been an important product since the middle of the 18th Century. It is a yellowish or greenish liquid with a typical odor and taste. It has been used in varnishes, histology, lithographic work and veterinary medicine.

Misc. Turpentines

Turpentines of small importance include Bordeaux turpentine from *Pinus pinaster*, Strasbourg turpentine from *Abies alba* and Jura turpentine from *Picea abies*. In he first two of these European turpentines, the crude exudation is used to some extent, but generally it is strained and filtered through cloth. The residue, known as Burgundy pitch, is a stimulant and counterirritant, and has been used in plasters, ointments and other pharmaceutical preparations.

Balsams

These are technically oleoresins that contain benzole or cinnamic acid and therefore are highly aromatic. However, the term "balsam" is frequently applied in error to much different substances, such as

Canada balsam, copaiba balsam, etc. The true balsams contain much less oil than the turpentines and are more viscous. They yield essential oils on distillation. Balsams are used in medicines and as fixatives in the perfume industry.

Balsam of Peru

This balsam is from *Myroxylon pereirae,* a tall tree in Central America. It is cultivated in many tropical and subtropical areas. The wood resembles mahogany and is valuable. The balsam is a dark, reddish-brown, thick, syrupy, viscous liquid, and is a pathological product formed as a result of wounding the tree. The tree trunks are beaten with a blunt instrument and the injured bark is then charred. It falls off naturally or is removed. Sometimes small "windows" are cut in the bark. The balsam exudes from the exposed surface of the wood and is collected on cloths.

These cloths are later boiled to free the purified balsam. Balsam of Peru has been used in medicine for healing wounds and skin diseases. During World War II it was used in field dressings for quick applications of a protective covering to the surface of wounds. It has also been used in the treatment of coughs, bronchitis and similar ailments, because of its simulative and antiseptic effect on the mucous membranes. It is utilized in perfumes as a fixative for the heavier odors and has served as a substitute for vanilla. Its common name is a misnomer because the tree does not grow in Peru. The main supply for the United States has been El Salvador.

Balsam of Tolu

This balsam is a pathological product obtained from *Myroxylon balsamum,* a tree of Venezuela, Colombia and Peru. Incisions with a V-shape are made in the trunk and the balsam exudes slowly and is collected in gourds. It is a brown or yellowish-brown plastic substance with a pleasant aromatic taste and odor. It has been used for salves and ointments and as an expectorant and antiseptic in the treatment of coughs, colds and bronchitis. It is sometimes used to flavour cough syrups. Considerable amounts have been used as fixatives in the perfume and soap industries.

Styrax

Styrax, or storax, occurs in two forms. The most important type, the styrax of antiquity, is known as Levant styrax. This is obtained from *Liquidambar orientalis,* a small tree found along the coasts of southwestern Asia Minor. The balsam issues as a pathological product from a wound. The outer bark is bruised and soon the balsam exudes into the inner bark. The outer layers are discarded and the balsam is recovered by boiling the inner layers in seawater.

The residual bark is dried and used for fumigation. Styrax is a semi liquid, grayish-brown, sticky, opaque substance with a pronounced aromatic odor. It has been used in soaps and cosmetics; as a fixative for the heavy oriental type of perfume; in adhesives, lacquers, and incense; as a flavouring for tobacco; and in medicine as a stimulant to the mucous membranes and for the treatment of scabies.

American styrax is obtained from the sweet gum, *Liquidambar styraciflua*.

Benzoin

This is a solid balsam that is often classified as a resin or balsamic resin. It is the pathological product obtained from several species of *Styrax* in Southeastern Asia and the East Indies. The balsam oozes from incisions made in the trunk and branches. Two varieties are known in commerce. Siam benzoin is from *Styrax tonkinense* and *S. benzoides* and occurs as yellowish or brownish pebble like tears with a milky white centre.

The tears are hard and brittle at room temperature and occur separately or adhere together slightly. Sumatra benzoin, from Styrax benzoin, occurs in reddish or grayish-brown blocks or lumps, composed of masses of tears stuck together. Benzoin is exceedingly aromatic with a vanilla like odor. It is used in medicine as a stimulant and expectorant; and in the preparation of heavy sweet perfumes, soap, toilet waters, lotions, tooth powders, incense and fumigating materials.

Other Oleoresins

Oleoresins that do not belong to either the turpentines or balsams are such products as copaiba, elemi and Mecca balsam.

Copaiba

Also known as copaiba balsam or copaiba, this product is a natural oleoresin obtained from several species of *Copaifera* from tropical South America. The copaiba trees are small with strong, tough, durable wood, which contains resin ducts of a large size. So much oleoresin is secreted and stored in cavities that sometimes the pressure causes the trunks to burst. Copaiba is obtained by boring holes into the heartwood. The secretion flows out very rapidly. It is a thin, clear, colourless liquid at first, but turns yellow and becomes more viscid with age. It has a peculiar aromatic odor and a persistent bitter taste. Several commercial varieties occur, which differ in the amount of resins and essential oils present as well as inn the source.

The most important are Maracaibo copaiba from Venezuela, obtained from *Copaifera officinalis*, which contains much resin and is rather thick; and Para copaiba, a very fine grade from Brazil, obtained from *C. reticulata*. Copaiba has been used in making varnishes, lacquers, and tracing paper, as a fixative in scenting soaps and perfumes, and in photography to emphasize half tones and shadows. In medicine it was used as a disinfectant, laxative, diuretic, and mild stimulant. Amerindians used it for its healing properties. It became known in Europe early in the 16th Century.

Substitutes for copaiba are Gurjun Balsam and Illurin Balsam.

Gurjun Balsam is obtained from *Dipterocarpus turbinatus* and related species from India and the East Indies. The thick, opaque, grayish oleoresin is obtained by cutting holes in the trunk.

Illurin Balsam or African Copaiba is a very fragrant oleoresin with a thick pungent, pepperlike odor that originates from a West African tree, *Daniella thurifera*. This species is sometimes called Sierra Leone Frankincense. It is found in the drier open forests and savannahs of Upper Guinea. The resin exudes in a copious flow from the base of the trees. A similar product comes from *D. oliveri*, a species of the denser moist forests. This oleoresin, under the name of wood oil, was once one of the chief products of Nigeria.

Elemi

This name is used as a collective for several oleoresins of different origin. It is also used in error as a synonym for some of the soft copals.

Elemis differ considerably in their characteristics. These oleoresins exude as clear pale liquids, but they tend to harden on exposure. Some remain soft, while others become very hard.

Manila Elemi is the most important and the best know of the elemis. Its source is the pili tree, *Canarium luzonicum*, of the Philippines. The oleoresin is secreted in the bark and oozes from the trunk in fragrant white masses. Native people have used elemi for torches and for caulking their boats.

Manila elemi has been used in lithographic work and in the manufacture of inks, adhesives, and cements; in the varnish industry to give toughness and elasticity to the products; in perfumes; and in medicine in plasters and ointments.

Of lesser importance is African Elemi from *Boswellia frereana*; Mexican Elemi from *Amyris elemifera, A. balsamifera* and other species; and Brazilian Elemi from *Protium heptaphyllum*, *Bursera gummifera* and other trees.

Mecca Balsam

This is a greenish turbid oleoresin with the odor of rosemary. It is obtained from Commiphora opobalsamum, in Arabia. This material has long been used in incense and for perfumes of an oriental type. It has some medicinal value. The supply of Mecca balsam has always been limited and thus is a rare and costly product.

Gum Resins

Gum resins are mixtures of both true gums and resins and naturally combine the traits of both groups. They often contain small amounts of essential oils and traces of colouring material. Gum resins occur naturally as milky exudates and collect in the form of tears or irregular masses. Injuring or tapping the plants also produces them.

They are mostly from plants of dry, arid regions, especially species of the Umbelliferae and Burseraceae. Three of the umbelliferous species, the sources of ammoniacum, asafetida and galbanum, are very common in Iran and Afghanistan and furnish the characteristic aspects of the vegetation of the plains and steppes in those regions.

During the dry season these plains are barren, but shortly after the rains begin, these plants send up thick stems from their perennial rootstalks. When fully grown the plants attain a height of 5-6 ft., and are so abundant that they from a kind of open forest.

Ammoniacum

This is obtained from *Dorema ammonniacum*, a tall, stout naked, hollow-stemmed perennial found in the deserts of Iran, southern Siberia and other parts of Western Asia. It has a milky juice, which exudes from the stem and flowering branches and hardens into tears. Insect injury often causes the exudation to occur.

The brownish-yellow tears are hard and brittle and occur singly or in masses. Ammoniacum is used in medicine as a circulatory stimulant, and in perfumery.

Asafetida

This is derived from Ferula assafoetida and allied species, stout perennial herbs of Iran and Afghanistan. The cortex of the thick fleshy roots exudes a milky juice during the rainy season. The crown of the roots is cut off and protected from the sun.

The gum resin gradually collects on the surface in the form of tears, or masses of tears of varying colours imbedded in a thick, gummy, grayish or reddish matrix. Asafetida has a powerful and foul odor and a bitter acrid taste, due to sulfur compounds present in the essential oil.

Despite this it has been used throughout the East for flavouring sauces, curries and other foods, and as a drug. In Europe and America is used in perfumes and for flavouring only when exceedingly dilute and after certain impurities have been removed. Asafetida has may valuable medicinal properties and is used in the treatment of coughs, asthma, and other nervous disorders and as an aid to digestion and metabolism.

Galbanum

Galbanum is a gum resin excreted from the lower part of the stems of *Ferula galbaniflua*, a stout herbaceous perennial of Northwestern Asia. It occurs in the form of separate tears or brownish and yellowish-green masses. It has been used for centuries in medicine. Galbanum has a powerful tenacious aromatic odor.

Myrrh

This is one of the oldest and most valuable of the gum resins. There has been some confusion regarding its source because two forms seem to occur.

Herabol Myrrh is derived from *Commiphora myrrha*, a large shrub or small tree of Ethiopia, Somaliland and Arabia. The gum resin oozes naturally from the stems or as a result of wounding.

The pale-yellow liquid gradually solidifies and becomes brown or even black in colour.

Herabol myrrh is used in perfume manufacture and for medicinal purposes as a tonic, stimulant, and antiseptic and is often a constituent of mouthwashes and dentifrices.

Bisabol Myrrh or Sweet Myrrh is from *Commiphora erythraea*, and Arabian species of similar appearance. This is the myrrh of antiquity and has been used for centuries in incense, perfumes and embalming.

Myrrh was an important product in Biblical times, ranking with gold in value. It is still used in perfumes and incense for religious ceremonies. It is one of the constituents of Chinese joss sticks.

Frankincense

This is also known as Olibanum and is obtained from *Boswellia carteri* and related Asiatic and African species. The clear yellow resin exudes from incisions made in the bark and hardens as small yellow grains.

Frankincense, like myrrh, has been a valuable material since Biblical times. It is still an indispensable ingredient of incense for religious purposes, and is also used in perfumes, because of its excellent fixative properties, face powders, pastilles and fumigating powders.

Opopanax / Bdellium

These gum resins have been of much less importance. Opopanax is derived from two very distinct plants, *Commiphora kataf* of the Burseraceae and *Opopanax chironium* of the Umbelliferae. It is used in perfumery and was formerly of importance in medicine.

Bdellium is a bitter aromatic gum resin obtained from *Commiphora mukul* of India and *C. africana* of Africa. It has been used in the perfume industry.

Misc. Resins

There are other resinous substances, which find no use in industry but are treated under "Medicinal Plants."

These include aloes, guaiacum, jalap and podophyllum, and will be treated later. Gamboge, a gum resin is discussed under "Dye Plants."

3

Drugs from Plants

Medicinal Plants

Plants have been used from ancient times to attempt cures for diseases and to relive physical suffering. Ancient peoples all had acquired some knowledge of medicinal plants. Oftentimes these primitive attempts at medicine were based on superstition and speculation. Evil spirits in the body were thought to be the cause of medical problems. They could be driven out of the body through the use of poisonous or disagreeable plant substances that rendered the body a disagreeable habitat. Medicine men or women of a tribe were usually charged with knowledge of such plants.

The progress of medicine has often been guided by the earlier observations and beliefs. Drug plants were always of especial interest. As early as 5,000 B.C. many drugs were in use in China. Sanskrit writings testify to methods of gathering and preparing drugs in these early times.

The Babylonians, ancient Hebrews and Assyrians were all familiar with medicinal plants. From Egypt there are records dating to 1,600 B.C. naming many of the medicinal plants used by physicians of that period, among which myrrh, opium, cannabis, aloes, cassia and hemlock are prominent.

The Greeks were familiar with many of the drugs of today, evidenced by the works of Hippocrates, Theophrastus, Aristotle and Pythagoras. The supernatural element continued to remain prominent in their culture, however. Only a few individuals were thought able because of some special power to distinguish harmful from valuable plants. This "rhizotomoi" or root diggers were an important caste in ancient Greece. In Rome there was less interest in plants that had healing powers. But by 77 BC Dioscorides wrote in his treatise, "De Materia Medica," dealing with the nature and properties of all the medicinal substances known at that time. This work was highly esteemed for 15 centuries and to this day is valued in parts of Turkey and North Africa. Pliny and Galen also described the nature of some drug plants (Hill 1952).

Following the Dark Ages there began a period of the encyclopedists and herbalists. The monasteries of Northern Europe produced large compendiums of information regarding plants, much of which was false. They stressed the medicinal value and folklore of plants. About the same time there appeared a "Doctrine of Signatures." This superstitious doctrine suggested that all plants possessed some sign, given by the Creator, which indicated the use for which they were intended. A plant with heart-shaped leaves was good for heart ailments; the liverleaf with its 3-lobed leaves was good for liver problems, etc. Many of the common names of plants owe their origin to this superstition. Names such as heartease, dogtooth violet, Solomon's seal and liverwort are examples.

Pharmacology and pharmacognosy owe their beginnings to the earlier beliefs and knowledge about medicinal plants. The interest in medicinal plants was especially pronounced among the early botanists who were often physicians.

Drug Plants

That branch of medical science dealing with the drug plants themselves is known as Pharmacognosy. It is concerned with the history, commerce, collection, selection, identification and preservation of crude drugs and raw materials. The action of drugs is Pharmacology. Worldwide there are several thousand plants that have been and are still being used for medical purposes. Many of these are restricted in use by native people who have long resided in any given area.

The Pure Food and Drug Act of 1906 in the United States has standardized most of the truly valuable drug plants. Such drugs are referred to as "official." Details about these plants may be found at the United States Pharmacopoeia, the Homeopathic Pharmacopoeia and the National Formulary, and various other sources in the United States and Europe. Very few drug plants are cultivated. Most of the drug supply is from wild plants growing in different parts of the world, especially in tropical areas. These drug plants are collected and prepared in a crude way for shipment. They eventually reach the centres of the drug trade and are processed. Sometimes a country has built up a monopoly of some particular drug.

For example, Japan used to control the export of camphor, agar and pyrethrum, while the Dutch in Java supplied almost all the Quinine (Chichona) for world trade. From 1920-1930 the importation of crude drugs increased 140 percent. Most of the processing of the crude material was carried out in the United States.

Additionally, several drugs are produced in the United States either from wild or cultivated sources. These include ginseng, goldenseal, digitalis, cascara, wormseed and hemp. When there are shortages additional plants grown are henbane, belladonna and stramonium.

A plant's medicinal value is due to the presence in its tissues of some chemical substance or substances that produce a physiological action on the body. Most important are the alkaloids, compounds of carbon, hydrogen, oxygen and nitrogen. Glucosides, essential oils, fatty oils, resins, mucilages, tannins and gums are all utilized. Some of these are powerful poisons so that the preparation and administering of them should be entirely supervised by physicians.

Drug Classification

Many methods have been proposed to classify drugs and drug plants. Classifications can be based on the chemical nature or the therapeutic value of the plant product, the natural affinities of the various species or the morphology of the plant organ from which the drug is obtained. Hill (1952) proposed a morphological basis of classification.

Generally it is found that the active principles are present in the storage organs of the plants, especially in roots and seeds, and less in leaves, bark, wood or other plant parts. The amount of the chemical substances present in any specific organ is so small that it is difficult to give any biological significance to it. There may be some slight protective function, but most likely the action that is valuable to humans in the treatment of disease are merely waste products of plant metabolism.

Drugs from Plant Roots

Aconite: This is obtained from the tuberous roots of the monkshood, *Aconitum napellus*. Although poisonous, its use in medicine is comparatively recent. The plant is a native of the Pyrenees, Alps and other mountainous regions of Asia and Europe. It is cultivated in temperate regions both as an ornamental and as a drug plant. Most of the commercial supply is from Europe. At first the leaves and flowering shoots were utilized, but later only the roots were used.

These are collected in the autumn and dried. Aconitine is the most important of the several alkaloids that are present. It is used externally for neuralgia and arthritis, and internally to relieve fever and pain.

Colchicum: Dried corms of the meadow saffron, *Colchicum autumnale*, are the source of colchicum. It is a perennial tulip like herb of Europe and Northern Africa. It possesses an alkaloid, colchicine, which is used in the treatment of arthritis and gout. Fresh roots seeds are also used to some extent. Colchicine has the ability to double the chromosomes in genetics studies.

Gentian (Bitterroot): *Gentiana lutea* is a tall perennial herb with striking orange-yellow flowers. Gentian is common in the mountains of Central Europe. The rhizomes and roots are dug out in the fall, sliced and dried. They contain several glucosides that are valuable as a tonic for they can be used with iron salts.

Goldenseal: Goldenseal, *Hydrastis canadensis*, was formerly common in the woods of Eastern North America. The Amerindians and European settlers used it as a remedy. The plant has been cultivated in the Pacific Northwest and North Caroline for it was almost exterminated as a wild plant. The roots and rhizomes contain several alkaloids that may be used as a tonic and for the treatment of catarrh and other inflamed mucous membranes.

Ginseng: Ginseng has been used in China since ancient times, where it is used to cure an array of diseases. True ginseng, *Panax schinseng*, is a plant of Eastern Asia and this was once the only source of the drug.

However, the demand became so great that large quantities of American ginseng, *Panax quinquefolium*, have been grown in recent years, and by 2003 Wisconsin led the production. In America ginseng is used as a stimulant and stomachic.

Ipecac: These are small shrubby plants in humid forests of the Neotropics. Several species are the source of this well-known drug, but the main source consists of the dried rhizome and roots of *Cephaelis ipecacuanha*. The main ingredient is Emetin, a white, bitter, colourless alkaloid. Ipecac is used as a diaphoretic emetic and expectorant. It is valuable in the treatment of amoebic dysentery and pyorrhea.

Jalap: This is a resinous drug obtained from the tubers of *Exogonium purya*, a twining, morning glory-like vine of the woodlands of eastern Mexico. The plant has been cultivated in Mexico, Jamaica and India. The roots are collected and dried over fires. Jalap is used as a purgative.

Licorice: This is a product that is known from ancient times. The plant, *Glycyrrhiza glabra*, is a perennial herb that grows wild in Southern Europe and Western and Central Asia. It is also cultivated in many parts of the same area. The largest producer of cultivated licorice root has been Spain. The roots are dried in sheds for several months and are shipped in cylindrical pieces. Licorice is used in medicine as a demulcent and expectorant and to disguise the flavour of medicinal preparations.

However, most of the supply is used as a flavouring material in the tobacco and candy industries and in the manufacture of shore polish. There are also many other industrial uses for licorice. It has a compound, glycyrrhizin, that is 50-times sweeter than sugar. A solution of it can be used for etching steel sections in photomicrographic work; and a substance is made from the waste root that foams readily and is used by brewers to give a head to beer. The fibres can be made into wallboard and boxboard under the name of "Maftex". It gives an insulating material and is made into Jacquard cards that are used for controlling the designs in the weaving of tapestries and other figured materials.

Podophyllum: Roots of the Mandrake or May Apple, *Podophyllum pellatum*, yield the drug podophyllum, which has been used in rural eastern United States as an emetic and cathartic.

The commercial supply of this widespread plant is southern Appalachia where it is collected wild or cultivated. Roots are collected in the autumn or spring and are cut into cylindrical segments and carefully dried. They contain a resin that is the source of the cathartic. East Indian podophyllum is obtained from *Podopjhyllum emodi* from the Himalayas.

Rhubarb: Two native shrubs of China and Tibet are the sources of the drug form of rhubarb: *Rheum officinale* and *R. palmatum*. These plants are similar to the garden rhubarb but grow to a much larger size. They have been extensively cultivated in China. The rhizomes and roots are dug and cut into short pieces or slices. These are threaded on a string and dried in the sun or in kilns. Rhubarb is used as a tonic and laxative and for indigestion. East Indian rhubarb is from *Rheum emodi*.

Squills: The white variety of sea onion, *Urginea maritima*, is the source of squills. The plant is native of the seacoasts of the

Mediterranean and has come under cultivation. The bulbs are dug up and the outer scales removed. The fleshy inner scales are then sliced and dried. Several glucosides are present. The drug is used as an expectorant and stimulant. A red variety contains toxic substances that render it useful as a raticide.

Senega: Senega snakeroot or milkwort, *Polygala senega*, is a small herbaceous perennial of Eastern North America. It is the source of a glucosidal drug obtained from the dried roots. The common name was derived because Senega or Seneca Indians used the plant as a cure for snakebites. Senega is used as an expectorant, emetic and stimulant.

Valerian: Dried rhizomes and roots of the garden heliotrope, *Valeriana officinalis*, furnish valerian. Native to Eurasia, it has long been cultivated in the United States as an ornamental. It contains an essential oil that is used to relieve nervous afflictions, pain, coughing and hysteria.

Drugs from the Bark of Plants

Cascara: Of North American origin, cascara is obtained from the reddish-brown bark of the western buckthorn, *Rhammus purshiana*, a tree of the northwestern United States and southwestern Canada. It was used by the Amerindians of the west and by the pioneer Spanish settlers who called it cascara sagrada, ors sacred bark. The bark is peeled in long strips during the summer and dried on racks. It is stored for a year before being used as a tonic and laxative.

Curare: Amerindians of Northern South America have long used a variety of poisonous extracts from various woody lianas as poisons. The identification of the constituent plant materials in curare is difficult because the sources vary from place to place. *Strychnas toxifera, Chondodendron tomentosum* and species of *Abuta* and *Cocculus*, as well as other species, have been used. New sources appear with continued explorations in the vast region. In the preparation of curare portions of the bark, roots, stems and tendrils are boiled down, the impurities skimmed off and the residue filtered. Catalytic agents are added and the whole mass is boiled to syrup. This is exposed to the sun and dried to a paste that is kept in tightly covered gourds or bamboo tubes.

Curare can cause progressive paralysis and eventual cardiac failure. These lethal effects are due to several alkaloids. One of these, curarine, has now been available to medicine for use in shock therapy, and as an ideal muscle relaxant. Curarine is also used for chronic spastic conditions, in surgical operations and tetanus and as a powerful sedative.

Quinine: One of the most important of all drugs, quinine has been a boon to mankind because it is the only adequate cure for malaria. Although some synthetic products are available, they only complement quinine and are not substitutes for it. Quinine is obtained from the hard thick bark of several species of the genus *Cinchona*, evergreeen trees of the Andes of South America. *Chinchona calisaya, C. officinalis, C. ledgeriana* and *C. succccirubra* are the principal species that have been used.

The Amerindians were familiar with chinchona bark. The first use of the drug by Europeans was in 1638 when, according to tradition, the Countess of Cinchon, wife of the Viceroy of Peru, was cured of malaria after all other remedies failed (Hill 1952). Although a story of doubtful accuracy, nevertheless the Jesuits were familiar with the use of the bark and carried it with them in their world travels. Very soon Jesuit's bark or Peruvian bark was in great demand. The supply at first seemed inexhaustible but diminished under the wasteful methods of collection. The trees were felled and the bark stripped off and dried in the open or over fires in huts. It soon became evident that only cultivation would guarantee the supply. Both the Dutch and the English sent collectors to South America, but the Andean natives guarded the remnants of the cinchona forests with great zeal.

However, despite the hostility a few seedlings and seeds were finally brought out of the area and became the basis of the great plantations of Java and India. Few tropical crops have been studied more intensively. All phases of cinchona production: breeding, culture, harvesting and processing were investigated. Eventually the Dutch developed a virtual monopoly, producing 95 percent of the world's supply. The Amerindian output was reduced to local consumption.

As the amount of bark produced was regulated in order to maintain high prices, attempts were made beginning in 1934 to establish a cinchona industry in the Western Hemisphere. An experimental plantation was begun in Guatemala and by the beginning of World War II a substantial mass of data and a nursery of superior clones from all parts of the world was available.

When the East Indian supplies were cut off because of the war, the United States instituted an extensive programme of cinchona procurement in the Neotropics, utilizing all available wild stands and developing new plantations. Several promising new sources were discovered, among them *Cinchona pitayensis*, a species that gave very high yields. From 1942-1945 exports of cinchona and quinine from the

Neotropics increased from 207,000 to over seven million pounds, with Ecuador and Peru being the main producers.

Cinchona bark is removed from cultivated trees by uprooting them when they are about 12 years old and stripping off the bark from both the stems and roots or by cutting the trunks above ground and stripping the felled portion. In the latter case adventitious roots develop and later the bark is removed from these in long quills. The most important constituent of cinchona bark is quinine. This is a very bitter, white, granular substance. In addition to its use in the treatment of malaria, it is valuable as a tonic and antiseptic and in the treatment of fevers. Over 29 other alkaloids have been isolated from the bark, including cinchonidine, cinchonine, and quinidine. All of these are useful in medicine. Totaquina is a standard mixture of all of these alkaloids.

Slippery Elm: The inner bark of the slippery elm, *Ulmus rubra*, is the source of this drug. The bark of this Eastern North American tree is removed in the spring and the outer layers are discarded while the inner portion is dried.

The bark has a very characteristic odor. It contains mucilage and is used for its soothing effect on inflamed tissues, either in the crude state or in the form of lozenges.

Drugs from Stems and Woods

Ephedrine

This is an alkaloid from the Asiatic *Ephedra sinica. E. equisetina* and other species of the same genus. These shrubs are low growing, dioecious, leafless with slender green stems.

The drug is extracted from the entire woody plant. In China the drug is known as "ma huang" for 5,000 years. In modern times it has been used in the treatment of colds, asthma, hay fever and other medical purposes.

Guiacum

This is a hard resin that exudes naturally from the stems of the lignum vitae trees, *Guaiacum offininale* and *G. sanctum*. It hardens into round, glassy greenish-brown tears. It is acquired from incisions, from the cut ends of logs or from pieces of the wood. Gum guaiac is used as a stimulant and laxative. It is also a good indicator of oxygen in the air. Lignum vitae, or Ironwood, trees are evergreens native to the West Indies and other Neotropical regions.

Quassia

There are two different sources for quassia. Jamaican quassia, *Picrasma excelsa,* is a tall tree of the West Indies and Surinam quassia, *Quassia amara,* grows in the Neotropics and West Indies. The latter is also a valuable timber tree with lustrous, yellowish-white fine-grained wood. Quassia is transported in the form of billets, and the drug is extracted by preparing an effusion of chips or shavings. It is very bitter to the taste and is used as a tonic and in the treatment of dyspepsia and malaria. It is also an insecticide.

Oleoresins: Please refer to Balsams and Oleoresins for additional medicines obtained from stems and wood.

Drugs from Plant Leaves

Aloe

Aloes are obtained from several different sources. Curacao or Barbados aloes are from *Aloe barbadensis* of the West Indies, Socotrine aloes from *Aloe perryi* of East Africa and Cape aloes from *A. ferox* of South Africa. These are tropical and subtropical fleshy plants with showy flowers.

The leaves contain a resinous juice with several glucosides. The juice slowly exudes from cut leaves placed in containers. It is evaporated in pans to a thick, viscous black mass that may be solidified. Aloes have been used as purgatives and as additions to skin salves. They seem to aid in the healing process of wounds.

Belladonna

Atropa belladonna is the source of this old and important drug. The dried leaves and tops and to some degree the roots contain the drug. The plant is a coarse perennial herb, native to Central and Southern Europe and Asia Minor. It is cultivated as a drug plant in the United States, India and Europe.

Leaves are collected during the flowering season and dried. They contain several alkaloids among which hyoscyamine and atropine are most important. Belladonna is used externally to relieve pain and internally to curb excessive perspiration and coughs. Atropine is used to dilate the pupil of the eye and as an antidote for organophosphorus insecticide poisoning.

Cocaine

Leaves of the coca shrub, *Erythroxylon coca,* and related species contain cocaine. Native to Bolivia and Peru, the plant is cultivated in

South America where the leaves are used as a masticatory. It is also grown in Sri Lanka, Taiwan and Java. The leaves mature in about four years when they are picked 3-4 times a year. They are carefully dried and shipped in bales.

The have a bitter aromatic taste. About 100 pounds of leaves yields one pound of the drug. Cocaine has been used as a local anesthetic and as a tonic for digestion and treatment of nervous conditions. It is addictive when used habitually. Some evidence suggests that cocaine was used in Ancient Egypt.

Buchu

The dried leaves of the shrubs *Barosma betulina*, *B. serratifolia* and *B. crenulata* contain the drug buchu. It grows in the dry mountainous parts of South Africa. The active ingredient is an essential oil that is used to disinfect and to stimulate excretion and also in the treatment of indigestion and urinary disorders.

Digitalis

Foxglove, *Digitalis purpurea* is native in Southern and Central Europe and has been used to treat disorders of the heart. The dried leaves are dried for use. It contains a glucoside, digitoxin. Its action improves the tone and rhythm of the heart beats thereby making contractions more powerful and complete. As a result more blood is sent from the heart, which aids circulation and improves body nutrition and hastens waste elimination.

Eucalyptus

The mature leaves of the blue gum, *Eucalyptus globulus*, contain an essential oil that is used in medicine. The tree can reach a height of 300 ft. in its native Australia. It was once extensively cultivated in California, Florida and the Mediterranean area. There was some belief that eucalyptus trees aid in eliminating malaria in countries where they are planted. Their extensive root system may play a role in drying-out mosquito breeding habitats. Eucalyptus oil is obtained from the dried leaves. It is used in the treatment of throat and nose disorders, malaria and other fevers. The colourless oil is yellow with a unique pungent odor.

Hamamelis (Witch Hazel)

Witch hazel, *Hamamelis virginiana*, is a shrub of Eastern North America. The dried leaves are gathered in southern Appalachia. However, in New England the bark, twigs and sometimes the entire plant are utilized. The active principle is tannin that is extracted with water and steam and distilled. Alcohol is added to the distillate in a ratio of about one part alcohol to seven parts distillate. Witch hazel is used as an astringent and to curtail bleeding.

Henbane

Henbane, *Hyoscyamus niger*, is a coarse smelly herb native to Europe and Asia. It has assumed weed status in other parts of the world. The drug is usually obtained in Europe, but during the World Wars the plant was cultivated in the United States. The leaves and flower tops contain several poisonous alkaloids: hyoscyamine and scopolamine. Henbane is used as a sedative and hypnotic. It acts in a similar manner to belladonna and stramonium, but is less powerful.

Hoarhound

Native to Central Asia and Europe hoarhound, *Marrubium vulgare*, has become naturalized in America where it is also cultivated. It is a small herbaceous perennial with white flowers in dense axillary whorls. Dried leaves and flower tops are used medicinally. Hoarhound is used as an infusion or in the form of candy or lozenges. It once was a favourite remedy for breaking up colds and has been used for arthritis, dyspepsia and other ailments.

Lobelia

The Indian tobacco, *Lobelia inflata*, is the source of this drug that is secured from the dried leaves and tops of wild or cultivated plants. It is a small North American annual with many blue flowers in leafy terminal racemes. It is also one of the few poisonous plants in North America. An alkaloid in lobelia is used as an expectorant, antispasmodic and emetic. Amerindians knew its properties. Some evidence suggests that tobacco was used in Ancient Egypt.

Pennyroyal

Pennyroyal, *Hedeoma pulegioides*, is a small aromatic annual found in poor soil in the eastern United States. An essential oil that it contains is derived from the dried leaves and tops of the plant. It has had some use in internal medicine. It has been an ingredient in liniments because of a counterirritant action. However, its main use was as an insect repellent.

Senna

Senna is an ancient drug that is obtained from dried leaflets and pods of several species of *Cassia* that are indigenous to arid regions in

Egypt and Arabia. Alexandrian senna is from *Cassia acutifolia* and East Indian or *Tinnavelly senna* is from *C. angustifolia*. Both species are cultivated in India. Leaves are picked, dried in the sun and baled. Senna is used as a purgative.

Stramonium

Thorn apple or Jimson weed, *Datura stramonium*, is the source of stramonium. The plant is highly poisonous and occurs worldwide although its origin was thought to be in Asia. However, Amerindians knew of its narcotic properties. It has been cultivated in Europe and the United States.

The drug is extracted from the dried leaves and flowering tops. The active principles are alkaloids that include hyoscyamine, atropine and scopolamine. The drug has been used as a substitute for belladonna for relaxing the bronchial muscles in asthma treatment. It has also been used in Asia for its narcotic effects.

Wormwood

A perennial plant of Northern Asia, Northern Africa and Europe, wormwood, Artemisia absinthium, is the source of an essential oil obtained by steam distillation from dried leaves and tops of the plant. The greenish liquid has been used in liniments. Over dosage can result in deleterious consequences. Its principal use is to flavour the liqueur absinthe, the use of which is prohibited in some countries. Absinthe contains other aromatics as well as wormwood. The plant has been grown in Oregon and Michigan.

Drugs from Flowers

Chamomile

Chamomile, *Matricaria chamomilla*, is a Eurasian daisy like plant that has become cultivated in many places. The dried flower heads contain an essential oil infusions of which are used as tonics and gastric stimulants. The flower heads of the Russian or garden chamomile, Anthemis nobilis, are used for similar purposes but also in poultices for bruises, sprains and arthritis.

Hops

The hop plant, *Humulus lupulus*, is native of the north temperate regions of Eurasia and America. The plant was known to the Romans and has been grown in parts of Europe since the 9th Century (Hill 1952). Hops are extensively cultivated in the United States, Europe,

South America and Australia. It is a climbing herb with perennial roots. These send up several rough, weak, angular stems with deeply lobed leaves and dioecious flowers. The female flowers are found in scaly, cone like catkins that are covered with glandular hairs. They contain resin and various bitter, aromatic and narcotic principles, mainly lupulin. The hop plants are trained on poles or trellises. Harvest is in the early autumn.

The catkins are carefully picked out dried in kilns at 70 deg. Fahrenheit or lower. They are treated with sulfur and baled for shipment. Hops are used in medicine for their sedative and soporific properties and also as a tonic. They have been used also in poultices. However, their principal use is in the brewing industry. Hops are added to beer to prevent bacterial action and decomposition, and also to improve the flavour.

Santonin

The Levant wormseed, *Artemisia cina,* contains a valuable drug known as santonin derived from the dried unopened flower heads. This is a small semi shrubby perennial of Western Asia. Most of the supply has come from Turkestan, although the species has been grown in the Northwestern United States. This drug is a good remedy for intestinal worms and has been used for this purpose for centuries. It was introduced into Europe during the Crusades.

Drugs from Fruits & Seeds

Chaulmoogra Oil

Natives of Burma and other parts of Southeastern Asia have used the seeds and oil from the chaulmoogra tree, *Hydnocarpus kurzii*, and related species to treat skin diseases. Thus, in a quest for a treatment for leprosy it was found at the University of Hawaii that the oil from these trees had certain acids the ethyl esters of which were productive in treating leprosy. These tall trees grow in dense jungles and bear velvety fruits with several large seeds. These contain fatty oil with a characteristic odor and acrid taste. The expressed oil is a brownish-yellow liquid or soft solid.

Colocynth

The bitter apple, *Citrullus colocynthis,* has a spongy pulp that when dried is the source of the glucosidal drug colocynth. The plant is native to warm parts of Africa and Asia, but has been distributed worldwide and cultivated in the Mediterranean area. The fruits resemble oranges, and the rind is removed while the white bitter pulp is dried and shipped in balls. It is a powerful purgative.

Cubebs

The dried unripe fruits of *Piper cubeda* are called cubebs. It is a climbing vine of Malaya and eastern India, and it is cultivated in Thailand, Java, Sri Lanka and the West Indies. The berries look a lot like black pepper, but they are stalked. They have a warm, butter aromatic flavour and a strong odor from the presence of an oleoresin. Cubebs are used to treat catarrh and as a kidney stimulant. They have also been used as a condiment or spice.

Croton Oil

The dried ripe seeds of *Croton tiglium* contains the fatty croton oil. It is a shrub or small tree of Southeastern Asia, but is also cultivated in Sri Lanka and India. Croton oil is a yellowish-brown liquid with a burning taste and offensive odor. It is one of the most powerful of purgatives. The flowers and crushed leaves are used in India to poison fish.

Nux Vomica: This is a valuable drug obtained from the seeds of *Strychnos nux-vomica,* a tree native to India, Sri Lanka, Cochin China and Australia. The large fruits contain from 3-5 grayish seeds that are very hard and bitter. Ripe seeds have two important alkaloids: strychnine and bucine. Nux vomica is used as a tonic and stimulant; strychnine is used in small doses to treat nervous disorders and paralysis. Its properties have been known back in the 16th Century.

Opium

One of the most useful and yet vicious drugs, opium is derived from the dried juice or latex of unripe capsules of the opium poppy, *Papaver somniferum.* The poppy is an annual with showy white flowers. Native to Western Asia, it is no found in most countries as a weed. It has been cultivated extensively in China, India, Asia Minor, the Balkans and elsewhere. Following petal fall the capsules are incised with a knife and the white latex exudes and soon hardens in the air. It is scraped off and shaped into balls or cakes, which are often wrapped in the poppy petals. Crude opium is a brownish material containing as many as 25 alkaloids, the most important and most powerful being morphine and codeine. Due to the narcotic and sedative action opium and its derivatives are used to relieve pain, relax spasms and induce sleep.

Psyllium

Commercial psyllium is the seed of several of the fleaworts, Asiatic and European species of plantain, which are cultivated in France, Spain and India. French psyllium is *Plantago indica*, Spanish psyllium is *P. psyllium*, and blonde psyllium, the East Indian product, is *P. ovata*. Psyllium seed contains a tasteless mucilaginous substance that acts as a mild laxative and is comparable to agar and mineral oil for use in chronic constipation. The extracted mucilage is used as a cosmetic and in stiffening fabrics.

Strophanthus

The dried ripe seeds of *Strophanthus kombe* and *S. hispidus* are the source of the drug strophanthus that is used as a heart stimulant. The plants are woody climbers of African forests. The active principles include the glucoside strophanthin and a few alkaloids. Another species, Strophanthus sarmentosus??, has a substance that can be transformed into cortisone.

Wormseed

American wormseed, *Chenopodium ambrosioides* var. *anthelminticum* is native to South and Central America and the West Indies, but has become naturalized in the United States. It has also been cultivated in many areas for its natural oil. The oil is obtained by distillation from the fruits and is used in the treatment of hookworm infections.

Drugs from Lower Plants

Antibiotics

These are substances produced mainly by certain harmless microorganisms that deter the growth and activity of various pathogenic bacteria. Antibiotics were not considered of importance until 1939 although they had been known previously. Since this time extensive investigations were carried out and a considerable number have been isolated and their therapeutic action studied. Molds, actinomycetes and bacteria are the chief sources, although antibiotics are also present in higher plants.

Penicillin

Best know of the antibiotics is penicillin. It was accidentally discovered in 1929 and reexamined in 1937. Soon it was recognized as an extremely valuable substance for combating staphylococcus, streptococcus and gas gangrene infections. It is acquired mainly from *Penicillium notatum*, a blue-green mold that occurs in floccose masses with a white margin. In gelatin substrate the mycelium excretes penicillin turning all to liquid. The crude penicillin is recovered, purified and dehydrated. It is an organic acid and readily forms salts and esters. Superior strains that yield greater quantities of the drug were developed. Other species of *Penicillium*, particularly *P. chrysogonum*, also produce the antibiotic. Penicillin is highly selective in its action and is effective against gram-positive bacteria. It is nontoxic and particularly useful in the treatment of bacterial endocarditis, gonorrhea, mastoiditis, local infections and certain types of pneumonia.

Streptomycin

Streptomyces griseus furnishes this antibiotic. It was first isolated in 1944 after a worldwide project testing soils. The organism is an actinomycete and is grown in deep submerged cultures. Streptomycin is especially effective against gram-negative bacteria and is used in the treatment of tularemia, empyema, urinary and local infections and some forms of tuberculosis, peritonitis, meningitis and pneumonia.

Aureomycin

Streptomyces aureofaciens, which was isolated in 1948 from soil, produces aureomycin. It is more versatile than penicillin or streptomycin by attacking not only gram-positive and gram-negative bacteria, but also the Rickettsiae, which had previously been immune to chemical assault. It has been used to combat forms of virus pneumonia, osteomyelitis, undulant fever, whooping cough and eye infections and where the patient has developed resistance to the other antibiotics or to sulfa drugs. Aureomycin is also a growth-producing substance.

Chloromycetin

This is a pure crystalline substance produced by *Streptomyces venezuelae*. It was isolated in 1948 after a search that involved the worldwide study of thousands of soil samples. It may also be produced synthetically. Chloromycetin, like aureomycin, is effective against the Rickettsiae. It is useful in the treatment of undulant fever, bacillary urinary infections, primary atypical pneumonia, typhus fever, typhoid fever, scrub typhus, Rocky Mountain spotted fever and parrot fever.

Terramycin

Terramycin is secreted by *Streptomyces rimosus* that was isolated from a piece of soil in Indiana after an exhaustive search involving many thousands of soil samples. It is valuable in treating common forms of pneumonia, typhoid fever, streptococcic and many intestinal and urinary tract infections. It is also effective against gram-positive and gram-negative bacteria, Rickettsiae and large viruses. It is somewhat different in therapeutic action from the other antibiotics.

Neomycin

This antibiotic that is produced by an organism resembling *Streptomyces fradiae*, has a complex composition and a wide range of experimental uses. It has been used to treat tuberculosis.

Misc. Antibiotics

There are many antibiotics known to be produced by bacteria. Among these gramicidin and tyrothricin from *Bacillus brevis*, bactracin and subtilin from *Bacillus subtilis* and polymixin from *Bacillus polymixa* have the best therapeutic possibilities.

Other Medicinal Substances

Agar

This is almost pure mucilage secured from various species of red algae. Japan used to be the principal producer of this product, utilizing *Gelideum corneum*, *Eucheuma spinosum*, *Gracilaria lichenoides* and other species found off the eastern coast of Asia. Some agar has been produced in the United States since 1919. However, during World War II production was greatly expanded in the United States. The principal species used were *Gelidium cartilagineum* on the Pacific Coast and *Gracilaria confervoides* on the Atlantic Coast. Agar industries have also been developed in Russia, South Africa and Australia, etc. The algae are collected, bleached and dried, and the mucilaginous material is extracted with water. Agar reaches the marked in flakes, granules or strips that are brittle when dry but become tough and resistant when moist.

The medicinal value of agar is in its absorptive and lubricating action. It is frequently used in a granular condition to prevent constipation. However, its greatest use is as a culture medium for bacteria and other fungi. In dentistry is has been valuable for making impressions for plates and molds. Cosmetic, silk and paper industries have found it valuable and is may also be used extensively as food.

Ergot

This is the dried fruiting body of a fungus, *Claviceps purpurea*, which is parasitic on rye and other grasses. The young fruit is attacked and when mature a purplish structure, the sclerotium, replaces the grain. Commercial ergot is chiefly from Europe where it is picked from rye plants or after the rye ahs been threshed by special machinery. Minnesota has also produced ergot. Wheat ergot is equally good as a drug. Ergot is used mainly to increase the blood pressure, especially in cases of hemorrhages following childbirth and other uterine disturbances.

Kelp

In Europe, the United States and Japan several of the larger brown algae have been used as a source of iodine, potash and other salts. In the United States the giant kelps of the Pacific involve mainly *Macrocystis pyrifera.* Kelp was also used as a source of acetone and kelp char, a bleaching carbon. There has also been attention given to the medicinal value of these seaweeds.

Other species, mainly *Laminaria digitata* and *L. saccharina* of the Atlantic and *Nereocystis luetkeana* of the Pacific, have been exploited as a source of algin, a valuable colloid extensively used in the drug, food and other industries. Algin or its salts, sodium alginate, is used as a suspending agent in compounding drugs; in lotions, emulsions and hand pomades; as a sizing for paper and textiles and in ice cream.

Lycopodium

Lycopodium clavatum and other club mosses contain about 50 percent fixed oils and so are but little affected by water. They are used as a covering for pills, as a diluent for insufflations and as a dusting powder for abraded surfaces. In industry they are used for making pattern molds and because of their inflammability, in flares, fireworks and tracer bullets. Europe and the northeastern United States have been the main producers.

Male Fern

Rhizomes and stalks of *Dryopteris felix-mas*, of North America and Eurasia, and the marginal shield fern, *Dryopteris marginalis*, of Eastern North America yield a drug known as male fern or aspidium. This is an oleoresinous substance that has been used for centuries for expelling tapeworms. The commercial supply ha been mainly from Europe.

Insecticides and Rodenticides

Pyrethrum

There are several sources for pyrethrum. Three of the most important species are Dalmatian insect flowers from *Chrysanthemum cinerariaefolium*, Persian insect flowers from *C. coccineum* and Caucasian insect flowers from *C. marshallii*. The Dalmatian species is favoured. It is a slender, glaucous, pubescent perennial 18-30 inches in height with pinnate leaves and small daisylike flowers.

It is a native of Dalmatia where it has been cultivated for centuries. Japan used to be the leading producer of pyrethrum flowers and they constituted one of its most valuable exports. Great care was exercised in gathering, drying and packing the crop. Later the species was being cultivated in California and other parts of the United States, Kenya, Italy, Australia, Brazil, Peru and Ecuador.

Pyrethrum is noninflammable, nonpoisonous and deposits no oily residue. It has been proven effective against flies, fleas, body lice and yellow fever and malarial mosquitoes. Pyrethrum bombs were standard equipment in malaria-infested areas. They contained the insecticide in a solvent under a pressure of 90 lbs. per sq.-in. A mechanical release allowed the vapour to escape through a valve in a fog. A 3-sec. application permanently paralyzed most insects. Pyrethrum coils are still in use in mosquito-infested areas. These are burned like incense and have a pleasant odor. Pyrethrum ointment is used in the treatment of scabies.

Rotenone

Rotenone-containing plants have poisons that were used by native peoples for centuries. The use of these climbers and creepers of the Leguminosae as fish poisons was noted by De Rochefort in 1665 and Aublet in 1775 (Hill 1952). Derris was in commercial use in the United States by 1911, but with variable and uncertain results. years of research resulted in standardizing the product and made it possible for widespread use.

This is a colourless crystalline compound together with related substances that occur as solids in the dried roots. The content may be as high as 12 percent. Rotenone is 15-times more toxic than nicotine and 25-times more than potassium ferrocyanide in killing insects. It has little or no effect on humans and other warm-blooded animals. Two principal sources are species of *Derris* in the Far East and *Lonchocarpus* in the Neotropics.

Derris

Natives in Malaya and Borneo for fishing have long used derris or tuba roots and arrow poisons. The various species of Derris are climbing vines typical of the jungle undergrowth from India to Indonesia and the Philippines. The plants have a short trunk, 3-4 ft. in height and 4 in. in diametre, with numerous long branches that climb over the vegetation.

The two most important species ore *Derris elliptica* and *D. trifoliata*. It may be propagated by cuttings and has been cultivated. Ecuador and Guatemala made it a commercial crop at one time. It grows well at low altitudes in deep, well-drained fertile soils. A dust made from the ground roots has marked insecticidal properties but it is nonpoisonous to humans, at least when taken through the mouth. The active ingredient, rotenone and a resin, may be extracted and used directly or in the form of soap.

Lonchocarpus

Roots of several species of *Lonchocarpus*, mainly *L. urucu* in Brazil, *L. utilis* in Peru and *L. nicon* in Guiana, make up an important source of rotenone. The plants are known also as cube, timbo and barbasco and are used by the Amerindians as fish poisons. At first they are bush like but later resemble vines and climb into trees. They thrive in the tropical forests at low altitudes where there is an 80-in rainfall and well-drained soil. At 2-3 years of age, the plant tops are cut away and the roots are dug up, dried, bundled and exported. They are then ground into a powder and mixed with talc or clay for dusting or with a liquid for spraying. Cube contains more rotenone than derris and, like derris, is an ideal insecticide for crop plants as there is no residue. Stem cuttings easily propagate Lonchocarpus, and there used to be many small commercial plantations. Cube first entered the world trade in 1934. Petroleum based alternatives have cut into this market.

Red Squill

This substance is obtained from the bulbs of the red variety of *Urginea maritima*, a native to the Mediterranean area. It is cultivated in Algeria. It is used as a raticide from ancient times and came into prominence again during the mid 20th Century. The toxic substance, a glucoside, has little effect on other animals.

Smoking & Chewing Products

Human beings have always chewed or smoked various substances for pleasure, for some physiological effect, in connection with religious ceremonies or in an effort to seek a flight from reality (Hill 1952). Although a few of these such as chewing gum are harmless, the majority of the materials that are smoked or chewed have a distinct stimulating or even narcotic effect.

This is due to the presence of various alkaloids. Betel, Cola and tobacco are less immediately harmful to the user than opium, cannabis and coca. The latter are true narcotics that contain alkaloids, which are detrimental even in small amounts. If used in greater quantities they may lead the addict to a low state of depravity and degradation by causing stupor, coma, convulsions and even death.

Tobacco

Tobacco, *Nicotiana tabacum*, has become an important commodity worldwide even as evidence mounts of the serious side effects on health through its carcinogenic properties. Tobacco is native to the Neotropical region, but the original wild ancestor has not been found. It has been under cultivation for centuries. The Amerindians originated the use of tobacco in connection with religious ceremonies. The plant had spread over North America in Pre-Columbian times. The first expedition of Columbus to America reported its use by the Amerindians. Tobacco was introduced into Europe in 1156 but at first was grown only for ornamental and medicinal values. Jean Nicot, for whom the genus was named, was responsible for its development in France.

From there it spread rapidly over the rest of Europe and into Africa, Asia and Australia. Smoking tobacco only became general after 1586, but ever since its popularity increased despite opposition from religious groups and governments that almost taxed tobacco products into extinction. Some evidence suggests that tobacco was used in Ancient Egypt.

The slight narcotic and soothing properties of tobacco come from the alkaloid nicotine. This active principle is absorbed by the mucous membranes of the nose and throat. Essential oils and other aromatic substances that develop during the curing and fermentation process cause the aroma and flavour.

Tobacco Types

Out of the some 50 species of tobacco known, only two have gained commercial importance. *Nicotiana tabacum* accounts for most of the production. This was the native tobacco of the West Indies and most of Central and South America. Originally a tropical plant, it has become adapted to cultivation in the temperate region also, and has been grown wherever climate permits.

There have been over 100 horticultural varieties developed. Tobacco is an attractive unbranched annual growing to a height of 3-6 ft. with large oval, usually sessile leaves with clasping bases. The branching influorescence bears pink flowers. The fruit is a capsule with many very tiny seeds. The leaves and stems are covered with glandular hairs that secrete a resinous fluid and are sticky to the touch.

Nicotiana rustica is a smaller and hardier species with yellow flowers. It is thought to have originated in Mexico and still grows wild in parts of North America. This species was cultivated and smoked by Amerindians in Eastern North America in Pre-Columbian times. It was the first tobacco to be grown in Virginia, but soon was replaced by *N. tabacum*. However, it continued to be produced in Central Europe, Northern Asia and the East Indies. This tobacco was used both for insecticidal and smoking purposes.

Tobacco Cultivation

Production of tobacco is highly specialized. The several types differ in their cultural requirements as to moisture, temperature, sunlight, soil and fertilizer. They also differ in methods of harvesting, curing and handling. The very tiny seeds are planted in seed beds and seedlings are transplanted when they are 4-6 in. high. A light sandy loam rich in humus and well fertilized with potash, lime and other essential elements is desirable. Continuous cultivation is necessary.

Tobacco that will be used as cigar wrappers is usually grown in the shade that insures a more uniform product. Once the plants have begun to grow, the terminal bud is removed to direct strength to the leaves. Frequently inferior leavs and suckers are also removed. When fully ripe, as indicated by a change in the colour of the leaves, either the entire plant is cut off or the leaves are harvested, one by one, as they mature in the case of cigar-wrapper production.

Tobacco Curing

The newly harvested plants or leaves are wilted and then are suspended in an inverted position from a framework in curing barns. There are two main methods for curing: air curing and flue curing. Fire- and sun curing are less common.

Air curing is a slow process carried out under mostly natural conditions in well-ventilated barns. The temperature and humidity must be carefully controlled. Artificial heat is applied only in unfavourable weather. Flue curing is a much faster process that is done in small barns with furnaces. Flue-cured tobacco develops a typical bright yellow colour, while fire curing involves drying over fires of charcoal or hardwood. They are virtually smoked without much increase in temperature. During this process they develop the odor of creosote.

This is the oldest method of curing and was practiced by the Amerindians. Sun curing is done in the open and used mainly in Turkey and the Orient.

Curing is actually an oxidation process or dry fermentation, during which the leaves lose most of their water and green colour and become tougher. Certain changes in chemical composition take place that are necessary for the development of the desired quality. Curing usually requires from 3-6 months. Freshly cured leaves are then sorted, fermented or "sweated," and aged before manufacturing. This task is generally accomplished in warehouses after the leaves have been graded. They are either piled up in large heaps or pressed into special containers. This phase may take from six months to three years, during which time the aroma and odor are developed. Undesirable traits are eliminated and the colour and burning qualities are enhanced.

Tobacco Grades

The proper grading of tobacco is a specialized task done by experienced personnel. A single crop may yield as many as 50 different grades. Various "classes" are recognized based on the method of curing and in the case of cigar tobacco on the use. The regions in which they are produced usually designate the types and the grades are based on use, texture, colour, quality and other traits. Seven common classes are: Class 1 (Flue cured); Class 2 (Flue cured); Class 3 (Air cured Light and Air cured Dark); Class 4 (Cigar Filler); Class 5 (Cigar Binder); Class 6 (Cigar Wrapper); Class 7 (Misc.).

Tobacco Uses

Tobacco may be used in several ways. The Amerindians used all snuff, chewing tobacco and smoking tobacco. Cigars and cigarettes were later developments. Throughout all the steps of manufacturing, especially in the case of cigarettes, it is necessary to maintain optimum moisture content. Adding various hygroscopic agents such as glycerin does this. These are called "humectants."

There is a great variety of flavouring and conditioning materials added to make chewing and smoking tobacco. These affect the taste and smoking qualities. They are utilized as a "sauce" in which the leaves are immersed or as a "spray." Among them are licorice paste, sugar, honey, molasses, rum and tonka beans. Deer's tongue, and old favoured flavouring, consists of the powdered leaves of *Trilisa odoratissima*, the wild vanilla of the southeastern United States. With tonka beans the flavour is due to the presence of coumarin.

Blending which involves the use of different grades of leaf is also an important feature in manufacturing. Perique and Latakia are common ingredients imported from Syria.

Snuff is made by grinding up dark air and fire cured leaves to a powder. The poorer grades and waste are often utilized.

Chewing tobacco is made from Burley, a dark air cured and flue-cured tobacco. It requires leaves that are rich in flavour, tough, gummy and highly absorptive to the various flavouring materials that are added. It was an early development of the industry, reaching its maximum production in the early 1900's. Navy plug is very sweet and thick and consists of filler with a wrapper.

Smoking tobacco is prepared from heavily sauced blends of Burley, flue cured and other tobaccos or from mildly flavoured straight Burley. Granulated tobacco, the oldest type, is blended whereas the plug cuts lacking wrappers consist of Burley.

Cigars were manufactured before cigarettes and reached their height of consumption in 1930. Formerly made by hand, they are now machine made except for the most expensive brands. Three grades of tobacco are utilized: fillers, binders and wrappers, all of which are air cured. Tobacco for fillers must have a sweet pleasant flavour and burn evenly with a firm white ash. For wrappers leaves free from flavour are required. They must also be thin and elastic with small veins and uniform in colour. Individual leaves are picked for this.

Cigarettes require light coloured leaves that lack gummy substances and have been either air or flue cured. The most spectacular phase of the tobacco industry had been the development of the cigarette. The first cigarettes were made from straight Virginia, flue cured or Burley tobacco. Today properly aged leaves are used and the stems are removed by hand or by machine. The moisture content is then increased to from 18-20 percent and the various grades are blended. Although the actual formulas are trade secrets, generally cigarettes contain about 53 percent flue cured tobacco, 33 percent Burley and 10 percent oriental and 4 percent Maryland. The leaves are then run through a cutting machine where they are shredded and dried. During the process the "casing" consisting of licorice, sugar, glycerin and various flavourings is added. Cigarette paper is made from flax fibre.

Tobacco Production

Most production of tobacco has been in the United States. This crop was first grown in 1612 and was first exported in 1618 from

Jamestown. From the very beginning tobacco was the backbone of the Virginia colony and even served as currency. A culture grew up around its cultivation in tidewater Virginia which has never been equaled in America and which flourished for two centuries. Following the American Revolution the industry declined, owing to competition of other countries and to soil depletion. Gradually the industry moved westward from the Piedmont region of Virginia and North Carolina. Tobacco has been grown in New England to some degree from the earliest days, but the crop has been important only since 1795. Later specialization confined the industry to certain areas that were better suited to one kind of tobacco or the other. The crop was grown commercially in 21 different states, with Kentucky and North Carolina producing about 60 percent of the total.

Despite the large domestic production a considerable quantity of tobacco has been imported, principally oriental types for use in cigarettes. Other large tobacco producing countries included China, India, Russia, Indonesia, Brazil, Turkey, Italy and Japan. Some areas, such as Cuba, produce an exceptionally high quality tobacco.

Misc. Tobacco Products

Tobacco extracts from low grades of fire- and dark air-cured leaves has been exported for making chewing tobacco. *Nicotiana rustica* has been grown as the source of nicotine that is extracted for use as an insecticide. Rutin, ordinarily obtained from buckwheat, is also present in tobacco leaves Nicotine sulfate had been an important insecticide until the development of organophosphorous compounds.

Betel

The chewing of betel nuts, or areca nuts, has been and still is widespread especially in Asia. By 1952 over 400-million were estimated to use this product. The widespread occurrence of the habit indicates its antiquity. Herodotus first described it in 340 B.C. (Hill 1952). In India, where it is called Pan over 100-thousand tons of the nuts are used annually. There it plays an important role in the daily life of the inhabitants.

Betel nuts are the seeds of the betel nut palm, *Areca catechu*. It is native to Malaya but is extensively cultivated wherever the nuts are used. Chewing the betel nut can be quite a complex process. A simple and most usual method involves the use of only three ingredients: betel nuts, betel leaves and lime. Slices are cured if the nuts are not wholly ripe or ripe nuts are placed in the mouth. Then fresh leaves of the betel pepper, *Piper betle,* are coated with lime and chewed with the nuts.

This practices always follows dinner or is done as a breath freshener. It has not been shown to be harmful but may even aid in digestion. Often mixtures of the nuts with cloves, cinnamon, cardamom, nutmeg or other spices are utilized. Another way of chewing betel involves the addition of tobacco.

Cola

Seeds of the cola tree, *Cola nitida,* are known as cola or kola nuts. They are extensively used in many parts of tropical Africa for chewing. The tree is tall with a straight trunk that can reach a height of over 50 ft. It grows wild in the forests of tropical West Africa. It is also cultivated in this region and the adjacent Sudan, and has been introduced into Brazil, India, Jamaica and parts of tropical Asia.

The fruit consists of star-shaped follicles that contain eight hard, Plano convex, fleshy seeds with a reddish colour and the odor of roses. These are marketed fresh and are usually chewed directly, although powdered nuts may be used.

Cola has played an important role in the social, religious and commercial life of the Africa natives. It was first reported in the 12th Century. The nuts are in great demand and the natives will go to considerable trouble to obtain them. Although they have a bitter initial taste they leave a sweet taste in the mouth after chewing. Using this product results in a slight stimulation and temporary increase in physical capacity.

It is reported to lower hunger and fatigue. Cola is a very complex caffeine-containing product. It not only contains two percent caffeine, but an essential oil and a glucoside, kolanin, as well.

The stimulating effect of cola is in part due to the caffeine and in part to the kolanin that acts as a heart stimulant. Old cola nuts tend to lose their kolanin. Chewing cola nuts has no effect on consciousness and produces no known deleterious results.

True Narcotics

Narcotic plants contain alkaloids that are valuable in medicine when used in exceedingly small amounts. They are used to relieve pain, produce sleep and quiet anxiety. However, serious physiological effects can easily result if they are not used with the utmost discretion.

Narcotic drugs vary markedly in their effects on the human body. Cocaine and opium act as sedatives on mental activity and bring about a state of physical and mental comfort. A lowering, and even suspension, of emotion and perception, accompanies this. Sometimes complete suppression of consciousness results. Cannabis, peyote, fly agaric, caapi and the solanaceous narcotics cause cerebral excitation and cause hallucinations, visions and illusions. Their use causes intoxication and may be accompanied or followed by unconsciousness or other symptoms of abnormal brain function. Kavakava is a sleep-producing drug that also produces a hypnotic state.

Coca

Coca, *Erythroxylon coca*, is the source of the drug cocaine. Chewing or the whole or finely powdered leaves of the plant is an ancient custom among the Amerindians of the Andes and the western half of the Amazon Basin. The plant was highly esteemed by the Incas who used it as an emblem of royalty (Hill 1952). The use of coca gradually spread among the common people in South America and Pizarro found it in widespread use in 1553. Some evidence suggests that cocaine was used in Ancient Egypt.

The use of coca aids one to resist physical and mental fatigue and to work for long periods without food or drink. The average consumption is 25-50 grams daily. The chewing of coca is followed after a short period of rest by greatly stimulated activity. The narcotic acts directly on the central nervous system, causing immediate psychic exaltation to such an extent that the consumer is able to forget hunger or other pain. It is habit forming and may lead to physical deterioration, sickness and even death as it favours malnutrition. The leaves are chewed with lime and the highly alkaline ashes of some plant, such as quinoa or cecropia leaves. The plant is widely cultivated on the eastern slopes of the Andes from Colombia to Argentina. It has also been cultivated in Java and India.

Opium

Opium is an ancient narcotic that is the dried juice that exudes from injured capsules of the opium poppy, *Papaver somniferum*. Its origin is probably Asia Minor, but it has spread to the West. The Greeks, Romans and Egyptians and probably the early Lake Dwellers of Switzerland knew the drug. It had also reached Iran, India and China by the 8th Century. When properly utilized, opium and the alkaloids derived from it are valuable medicinally and have proved a blessing to humans in the relief of pain. Excessive use of the drug and the resulting opium addiction has been and continue to be the cause of unbelievable suffering. No other drug has caused so much corruption and tragedy.

In spite of every effort to stamp out the habit it has increased worldwide. By 1952 it was estimated that 900-million people were using opium as a drug.

In India opium has usually been eaten and the habit was common to all classes of society. So great has been the demand that the cultivation of the opium poppy continues as one of the most profitable industries in Pakistan and Afghanistan especially. During earlier times of the trade, the Dutch, Portuguese and English openly exploited opium. Later the various governments began strict regulation of its availability. However, the recent invasion of Afghanistan by Western nations has led to a breakdown in curtailing the production of opium poppies, so that by 2003 the crop is once again spreading among farmers in rural areas.

In China the usual method of consumption was opium smoking by placing a small pellet in the bowl of a special pipe and inhaling the fumes. In this way more morphine is said to be absorbed and the effects on the body may be greater.

In Europe and the United States, although opium, morphine, heroin, codeine and other derivatives are used as medicines, the smuggling and use of these narcotics are matters of utmost concern.

Cannabis

The hemp plant, *Cannabis sativa*, is the source of a textile fibre and a drying oil also yields a narcotic drug. The dried flowering tops of females plants, pressed together into solid masses, constitute the official "cannabis indica." The drug can be used in medicine to relieve pain and in the treatment of hysteria and various nervous disorders. The active principle is resinous in nature and contains 3-4 very powerful alkaloids. Hemp is frequently cultivated solely as a drug plant. This is especially true in India where hemp growing has developed to the level of a science. The use of hemp as a narcotic is old and extends back to 3,000 B.C., first in China and later in India. The plant was used by the Assyrians and was known to Herodotus.

Indian hemp is consumed in various ways. The pure, undiluted sticky yellow resin, which is naturally exuded from the flowering tops of cultivated female plants, is known as charas or hashish. Formerly the resin was obtained by rolling or treading on the leaves or by having natives run quickly through a mass of the plants. The resin stuck to the body or clothes of the runner and was subsequently removed.

Today it is carefully pressed out of the flowering tops between layers of cloth and then scraped off. Charas is smoked. It is the most powerful form of the drug. Bhang consists of the tops of wild plants, which have lower resin content in a water or milk mixture. It is also smoked.

In America this type of hemp is known as marijuana. Ganja is a specially cultivated and harvested grade of hemp used for smoking and in beverages and candies. It has high resin content. Some evidence suggests that marijuana was used in Ancient Egypt.

Cannabis in its different forms can produce serious side effects for the consumer. It causes a stupefying and hypnotic effect, accompanied by hallucinations, agreeable and often erotic dreams, and a general state of ecstasy. Addicts while under the influence of the drug may show emotions of happiness, be noisy or may even become fanatical and commit murder. However, when used in moderation it seems not to be harmful.

Peyote

The cactus, *Lophophora williamsii*, is the source of peyote or mescal buttons. It is indigenous to northern Mexico and the southwestern United States. It occurs on dry, arid plateaus in a limited area bordering on the Rio Grande.

This cactus has the appearances of a large carrot with all parts of the plant, except a button like top, below the ground. These tops are round and flattened and are wrinkled. They do not bear spines but little tufts of silky hairs. Some of the buttons are very tiny while others may reach over 3 in. in diametre. Peyote contains several powerful alkaloids with narcotic properties. Amerindians of the region have used peyote for hundreds of years during their religious ceremonies. Despite serious opposition the habit is actively practiced among some Amerindian groups that still maintain a sacred cult.

The buttons are chewed in either the fresh or dried state until they are soft. They are then rolled up in the hand into little pellets, which are swallowed. A beverage may also be prepared by boiling the buttons. Peyote produces a state of ease and well-being, accompanied by visions and hypnotic trances. Users find themselves in a world full of new sensations and pleasures. Peyote also has also been used extensively for its presumed medicinal value as it is supposed to cure bodily ills.

Fly Agaric

Fly agaric, *Amanita muscaria*, is known in Asia, America and Europe as one of the most poisonous fungi. Despite this quantities of this mushroom are consumed in Siberia and other parts of Northeastern Asia for its intoxicating effect. Dried agarics are chewed until soft, rolled into cylinder like pieces and swallowed. At other times they are used in milk, water or the juice of berries as a beverage.

Using this fungus produces hallucinations and illusions accompanied by giddiness, involuntary words and actions and finally unconsciousness. Two small fungi are enough to produce an intoxication that will last a whole day. Prolonged use of this narcotic will completely shatter the nervous system, yet cravings for the drug among its addicts is very great.

Ololiuqui

The native people of southwestern Mexico were using a narcotic plant in religious ceremonies and as an intoxicant in Pre-Columbian times. Ololiuqui is a decoction made from the seeds of *Rivea corymbosa*, a woody vine resembling morning glory.

It is still used primarily in Oaxaca in medicine as a means of divination and as a narcotic. Ololiuqui induces a hypnotic sleep or coma with hallucinations and a feeling of ecstasy, but with no unpleasant aftereffects.

Caapi

Amerindians living in the Amazon Basin utilize *Banisteriopsis caapi*, *B. inebrians*, *B. quilensis* and species of the genus *Tetrapterys* as the source of a narcotic beverage. These plants are called caapi in Brazil, ayahusca in Peru & Ecuador and yaje in Columbia. They are lianas that grow naturally in the virgin forests and are sometimes cultivated in native villages. To prepare the beverage the lower part of the stem is cut off, cleaned, macerated and boiled or utilized as a cold decoction. Caapi is used in religious ceremonies. It produces visions, dreams and other mental disorientations. It is also an excitant and induces courage.

Solanaceous Narcotics

Certain plants in the family Solanaceae contain alkaloids that produce disorders of the brain and excitation when smoked or consumed. These narcotics are often responsible for some of the incomprehensible acts of fanatics in the East and elsewhere.

The genus *Datura* has been and still is extensively used in all the continents except Australia for its narcotic and hypnotic properties. The Jimson weed, *Datura stramonium*, the source of the drug stramonium, was known as a narcotic as early as 37 A.D. (Hill 1952). It is still a favourite source for "knockout drops" in the tropics.

The maikoa, *D. arborea,* and *Datura sanguinea* are subtropical shrubs of South America that have been used by various groups of the westernmost Amazon Region for their narcotic properties. *The Aztecs in Mexico as a medicine and in their religious ceremonies used datura innoxia.* It is still used by some of the local inhabitants. Other species are also used elsewhere, all of them producing comparable effects, such as sense illusions and motor disturbances as well as senseless activities and loss of memory. At least one student at the University of California Riverside campus literally became a vegetable after having experimented with a local species of *Datura* growing in the area. This cost him his research assistantship and reduced him to a supervised livelihood of sweeping floors.

4

Essential Oils

Camphor

Camphor is an important essential that is used in industry. Commercial camphor, called camphor gum, consists of tough, white translucent masses or granules with a penetrating odor and pungent aromatic taste. It is solid at room temperature, thus bearing the same relation to the other essential oils that vegetable fats do to the fatty oils. It volatilizes very slowly.

The oil is obtained by distillation of the wood of the camphor tree, *Cinnamomum camphora,* and native to China, Taiwan and Japan. This is a very tall and striking tree with shiny, dark evergreen leaves. The tree has been widely introduced into tropical and subtropical regions, mainly as an ornamental plant. The camphor industry is centreed in Taiwan. Earlier crude methods of obtaining camphor were very destructive and the existence of the tree was threatened. Finally only trees 50 years of age or older were used and every stages in the process was carefully regulated.

The wood is reduced to chips or ground to a fine powder and the leaves are also ground up. This is then distilled with steam for several hours and the crude camphor crystallizes on the walls of the still. This is removed and must be purified before it is ready for market. Synthetic camphor from pinene, a turpentine derivative, gradually dominated the market.

The principal use of camphor has been in the manufacture of celluloid and various nitrocellulose compounds. It also has a wide range of medicinal uses, both internally and externally. It is also used in perfumery.

Borneo camphor, obtained from *Dryobalanops aromatica* of the east Indies, has been used as a substitute.

Cedarwood Oil

Cedarwood oil, along with clove and bergamot oils, is one of several of the essential oils have a high refractive index and are valuable as clearing agents in the preparation of permanent microscopic mounts and for use with oill-immersion lenses. This inexpensive oil is obtained by steam distillation from the heartwood of the eastern red cedar, *Juniperus virginiana*, and related species. Wood chips, sawdust, waste from the lead pencil and other industries, old stumps, roots and even fence rails have been utilized. Cedarwood oil is also used in perfumery, soaps, deodorants, liniments, cleaning and polishing preparations and as an adulterant of expensive sandalwood and geranium oils. It has insecticidal properties and is used as a moth repellent and in fly sprays.

Misc. Oils

Essential oils are also useful as solvents for paints and varnishes. The most important of these is oil of turpentine. Various other oils, mainly eucalyptus oil from *Eucalyptus dives*, are employed in the flotation process for the separation of minerals from their ores. Still other volatile oils have been used in the preparation of cleaning materials and other industrial purposes.

Fatty Oils

Fatty oils are fixed oils because they do not evaporate or become volatile. They cannot be distilled without being decomposed. Chemically fatty oils are close to animal fats. They contain glycerin in combination with a fatty acid. They are liquid at room temperature and usually contain oleic acid.

However, fats are solid at room temperature and contain palmitic or stearic acid. Fatty oils are insoluble in water but soluble in several organic solvents. Breakdown products of fats are fatty acids and glycerin accompanied by a rancid odor and taste. Boiling a fat with an alkali causes it to decompose and the fatty acid unites with the alkali to form soap. The addition of lye or potash will make the soap softer, if soda is used a hard soap results.

Fatty oils occur in many plant families, both tropical and temperate. They are stored, frequently in large amounts, in seeds and somewhat in fruits, tubers, stems and other plant organs. They may also contain proteins. This kind of reserve food material is available as a source of energy for the processes involved in seed germination. Fatty oils are bland and lack the strong taste, odor and antiseptic qualities of essential oils. Thus they are suitable for human food. These edible oils contain both solid and liquid fats and form an important part of the human diet.

Extraction of fatty oils varies. Usually the seed coats have to be removed and then the remainder is reduced to a fine meal. The oils are removed by solvents or by subjecting the meal to hydraulic pressure. The residue is rich in proteins and is valuable as an animal food and fertilizer. Pressure causes the cell walls to break and the fats are released. The extracted oils are filtered and may be further purified. Higher grades are edible and lower grades are used in various industries. Fatty oils may also have medicinal value.

Four classes of plant fatty oils are (1) drying oils, (2) semidrying oils, (3) nondrying oils, and (4) fats or tallows. The drying oils can absorb oxygen and on exposure dry into thin elastic films. These oils are importance in the paint and varnish industries. Semidrying oils absorb oxygen slowly and only in small amounts. They form a soft film only after long exposure to air. Some are edible; others are used as illuminants or in making soap and candles. The nondrying oils remain liquid at room temperature and do not form a film. Such oils are edible and may be used for soap and lubricants. The fats are solid or semisolid at room temperature. They are edible and also useful in the manufacture of soap and candles. Drying and semidrying oils are more common in plants of temperate climates, while nondrying oils and fats predominate in plants of tropical areas.

Drying Oils

Linseed Oil

Flax seed, *Linum usitatissimum*, has been the source of one of the most important of the drying oils. The oil content is 32-43 percent. The seeds are collected and stored for several months. Then impurities are removed and the seeds are ground to a fine meal. The oil is usually extracted by pressure with heat or by the use of solvents. Cold-pressed oil is produced in Eastern Europe where it is used for human consumption. Linseed oil varies from yellow to brownish in colour and has an acrid taste and smell. It forms a tough elastic film when oxidized. Heating the raw oil to 125 deg. Centigrade increases this drying property. This produces the "boiled" linseed oil. Linseed oil has been used mainly in making paints, varnishes, linoleum, soft soap and printer's ink. Following extraction, the oil cake can be used as an animal feed. Argentina has been the main producer, where over six million acres were devoted to seed-flax cultivation by the mid 1900's. Russia, India, China, Uruguay and Canada and the United States also produced considerably quantities. Montana, the Dakotas and Minnesota were the centres of seed-flax production in the United States, with an annual yield of 40 million bu. by 1950.

Tung Oil

Tung oil, or Chinawood oil, has been widely used in the varnish industry and as a substitute for linseed oil. It is obtained from the seeds of two Chinese species of *Aleurites, A. fordii*, the tung-oil tree, a species native to central and western China, and *A. montana*, the mu tree, found in southwestern China. Oils from these trees are almost identical in composition and properties and imported tung oil is often a mixture of the two.

In China tung oil has been used for centuries in waterproofing wood, paper and fabrics. It is a good preservative and is resistant to weathering, so it is especially valuable for painting outdoors. Boatmen have sought after tung oil because it is little affected by water.

In the United States cultivation of *A. fordii* began in 1905 and grew to 75,000 acres in the Gulf States by the mid 1950's. Tung trees are handsome and are often planted as ornamentals. The tung oil industry had assumed great importance in southern agriculture by the mid 1950's. Not only was it a profitable source of income, but also the orchards could be planted on otherwise useless eroded land.

The outer husk of the fruit is removed and the oil is expressed from the seeds by expeller presses. Tung oil is pale yellow to dark brown and dries very rapidly and also has preservative and waterproofing qualities. Consequently, its chief use has been in the varnish and paint industries, where it largely replaced kauri and other hard resins. Large quantities were also used in producing linoleum, oilcloth, brake linings, soap, leather dressings, inks, insulating compounds and fibreboard. The oil cake is a good fertilizer but is unsuitable as an animal feed.

Soybean Oil

Soybean, *Glycine max*, is native to China and has been a most important food plant in Eastern Asia. The oil is midway between linseed and cottonseed in its characteristics, so that sometimes it is classified as a drying oil or semidrying oil. The oil is extracted from the seeds by expression with hydraulic or expeller presses or by the use of solvents.

The oil content of improved varieties now exceeds 22 percent. After refining, soybean oil can be used in salads or as cooking oil and for other food purposes.

There are many food products made from it, such as margarine, tofu, ice cream substitutes, vegetarian meat substitutes, etc. It is also used in the manufacture of soap, candles, varnishes, lacquers, paints, linoleum, greases, rubber substitutes, cleaning compounds disinfectants and insecticides. The oil cake or meal has a 40-48 percent protein content and is valuable as a meat substitute and feed for livestock. It is also used for adhesives, plastics, foaming solutions, spreaders, fertilizers, sizings, synthetic textiles, etc.

Oiticica Oil

Seeds of large evergreen tree of northeast Brazil, *Licania rigida*, provide this oil. It has been used as a substitute for tung oil. It is extracted by solvents or by hydraulic presses. This oil finds use in the paint and varnish industries; in making linoleum, printing inks and brake bands; and for improving the elasticity of rubber. It has been used in Brazil as an illuminant and medicine.

Perilla Oil

The seeds of *Perilla frutescens*, an aromatic annual, 3-5 ft. in height with numerous branches, provides this oil. Native to northern India, China and Japan, it is extensively cultivated in the Orient, especially in Manchuria and Japan. The plant matures slowly and must be harvested before it is completely ripe or the seeds will fall from their capsules. The oil is expressed from roasted and crushed seeds and is edible, having been used for centuries. Industrial uses of perilla oil are much more important. It is deployed in the manufacture of Japanese oilpapers, cheap lacquer, paper umbrellas, waterproof clothes, printer's ink and leather. A considerable amount of this oil was imported into North America during the mid 1950's as a substitute for linseed oil.

Candlenut Oil

The oil is from the hard-shelled seeds of *Aleurites moluccana*, native to Malaya and Pacific Islands. In the Philippines it is called lumbang oil. It is a good drying oil and has been used in making paints, varnishes, lacquer, linoleum and soft soap and as a preservative for the hulls of ships. The nuts were at one time used in Hawaii and other Polynesian islands for illumination, thus the name candlenut. The oil cake is poisonous and may be used only as a fertilizer.

Walnut Oil

The mature and old kernels of English walnut, *Juglans regia*, and yield a drying oil used for white paint, artist paints, printing ink and soap. Hot-pressed oil is adapted for these purposes. The fresh oil and cold-pressed oil have a pleasant odor and nutty flavour and are edible. Waste kernels from the shelling process may be used as a source of the oil.

Niger Seed Oil

This oil is pale yellow and comes from the seeds of *Guizotia abyssinica*, an annual plant in tropical Africa. It is cultivated in India, Africa, Germany and the West Indies. Higher grades have a pleasant aromatic odor and are used for food, while poorer grades are made into soak or serve as illuminants. There has never been a widespread use of this oil in the Western Hemisphere.

Poppy Oil

Seeds of the poppy, Papaver somniferum, provides an important drying oil. This drug plant is grown for its edible seeds in northern France and Germany and in India. The first pressing yields white edible oil, while a second hot pressing furnishes reddish oil used for lamps, soap and, after bleaching, for oil paints.

Safflower Oil

Safflower, *Carthamus tinctorius*, is the source as a dye as well as oil from its seeds. It is extensively cultivated in Egypt, India and the Orient and somewhat in North America. It is widely used in Mexico Uses include manufacture of paints, soap, varnishes, illuminant and edible oil.

Tall Oil

This is really not true oil but a byproduct of the sulfate pulp industry. The waste liquor from pine pulp mills is concentrated by evaporation, and the soap curds are removed and acidified. Crude tall oil, which results, contains fatty acids, resins and other substances. It is refined by steam distillation. It is used in the manufacture of soap and, after treatment with glycerin, as a drying oil.

Other Drying Oils

Drying oils may be obtained from many other plant species, some of which have commercial importance. Among them is hemp, *Cannabis sativa*, grown in China, Japan and Europe and the oil is extracted from seeds. It is used for soap, paints, varnishes and lamp oil. The seeds of tobacco, *Nicotiana tabacum*, Hevea brasiliensis, *Manihot glazovii* and other sources of rubber furnish oil as well as grapes and raisins. Grape-seed oil in particular has become popular in California as a food product.

Semidrying Oils

Cottonseed Oil

The most important of the semidrying oils, cottonseed oil is used as the standard of comparison. The United States has been the main producer, but almost all the cotton-growing nations provide varying amounts. Over one billion pounds of the oil were expressed annually in the United States by the mid 1950's. The industry was first established around 1880, prior to which cottonseeds were merely discarded. The seeds are carefully cleaned and freed from impurities and the linters and usually the hulls are removed. The kernels are then crushed and heated and are finally subjected to hydraulic pressure or expeller presses. The oil is pumped into tanks where impurities settle out. The pure refined oil is of value as a salad and cooking oil and for making margarine and lard substitutes. The residue is the source of various products that have a wide range of industrial uses. Among these are soap, oilcloth, washing powders, artificial leather, roofing tar, insulating materials, putty, glycerin and nitroglycerin. Cottonseed meal is important as a food for animals and as a fertilizer.

Corn Oil

There is about 50 percent oil located in the embryo of maize kernels. It is used for a wide variety of purposes. Refined oil is for human consumption either directly or in margarines, while the crude oil has industrial uses such as the manufacture of rubber substitutes, paints, soaps. Corn oil has little lubricating value.

Sesame Oil

Also known as gingelly oil, sesame oil is the product of seeds of an annual herb, *Sesamum indicum*. It has been a chief oil of India and was cultivated there and in China from ancient times. Its use has spread to other tropical regions and it is now grown in many Asiatic, African and tropical American countries. China produces about one-half of the world supply, India one-third and Africa and America the rest. Slaves brought sesame oil to North America, and in the southern United States the plant was grown in slave days.

The seeds contain about 50 percent oil that is easily extracted by cold pressure. The finer grades are tasteless and nearly colourless and are used as a substitute for olive oil in cooking and in medicine. Enormous quantities of sesame oil are used in Europe in the manufacture of margarine and other foods. Poorer grades of oil are used for soap, perfumery and rubber substitutes and sometimes as a lubricant. In India the oil is used for anointing the body, as fuel for lamps and as food. Oil cake is a good cattle feed, and sesame seeds are also used in confectionery and baking industries.

Sunflower Oil

Common sunflower seeds, *Helianthus annuus,* contain 32-45 percent of light golden-yellow oil equal to olive oil in its medicinal and food value. It is an excellent salad oil and is used in margarines and lard substitutes. The seeds are a nutritious bird and poultry food, while the oil cake is excellent for livestock and the whole plant is often grown for silage. The oil has semidrying properties rendering it useful in the varnish, paint and soap industries. Its origins probably lie in South America but it is cultivated worldwide. Rumania, Russia and Argentina are big producers of sunflower seeds, while increasing acreage is planted in North America.

Rape and Colza Oils

Seeds of several species of *Brassica,* mainly *B. campestris, B. napus* (rape) and *B. rapa,* yield oils with characteristics that classify them as rape or colza oils. The oil content is 30-45 percent and expression or solvents extract the oil. Rapeseeds have been widely cultivated in Europe, China, Japan and India. The crude oil is edible when cold pressed and is used for greasing loaves of bread before baking. It is used in lamps, in oiling wooden goods, in the manufacture of soap and rubber substitutes and for quenching or tempering steel plates. The refined oil, or colza oil, is edible and it can be used as a lubricant for delicate machinery.

Other Semidrying Oils

Other sources are camelina oil from *Camelina sativa,* which is grown in Europe for its seed and used for soap and as an illuminant; croton oil, a powerful drug that is treated under "Medicinal Plants;" and argemone oil from *Argemone mexicana.* The seeds of many varieties of cultivated plants, such as pears, apricots, apples, peaches, cherries, plums, citrus, cereals, tomatoes, canteloupes, watermelons, pumpkins and black and white mustard also contain semidrying oils.

Nondrying Oils

Olive Oil

The fruits of the olive, *Olea europaea*, provide olive oil, and it is the most important of the nondrying oils. The tree is a small evergreen cultivated principally in Mediterranean countries and to a lesser degree in South Africa, Australia, South America, Mexico and the United States. The world production of olive oil has exceeded two billion pounds since the mid 1950's, with Spain, Italy, Greece and Portugal the leaders. Oils from the latter countries are believed superior in flavour to those from other areas.

The oil is squeezed from the pulp either by hand or mechanically. The finest grades are obtained by the former method. These oils are golden yellow, clear and limpid. They are used chiefly as salad and cooking oils, in canning sardines and in medicine. Inferior grades have a greenish tinge and are used for soap making and as lubricants. The poorest grades are obtained by the use of solvents after several pressings. Fully ripe olives give the largest yield. Olive oil is one of the most important food oils as it will store for long periods and becomes rancid only when exposed to the air. The oil cake is used for stock feed and as a substitute for humus in coil conditioning.

Peanut Oil

Seeds of the common peanut, *Arachis hypogaea*, originally from South America, provide peanut oil. The plant was widely used by Amerindians in Peru and so highly regarded that gold replicas are found in the tombs.

The main producers are China, India, Africa, and the United States. The seeds are shelled, cleaned and crushed, and both hydraulic presses and expellers express the oil. The filtered and refined oil is edible and used as a salad oil, for cooking, for packing sardines, in making margarine and shortenings, and as an adulterant for olive oil. Inferior grades are for soap making, lubricants, and illuminants. The oil cake is an excellent livestock feed as it has higher protein content than any other similar product. Peanut oil is in great demand in Europe where it is extracted with solvents as well as by expression. Spanish peanuts are grown in the United States for oil production because they have higher oil content.

Castor Oil

Seeds of *Ricinus communis* furnish versatile oil. The plant is a coarse erect annual herb that is cultivated in both temperate and tropical regions. In North America it is a favoured ornamental plant but has been grown for its oil. The characteristically marked seeds contain 35-55 percent of thick colourless or greenish oil that is obtained by expression or solvent extraction.

The chief use of castor oil used to be in medicine, where it acts as a purgative. Presently most of the production is utilized in industry in the manufacture of over 25 different products. It is water resistant and thus may be used for coating fabrics and for protective coverings for airplanes, insulation, food containers, firearms, etc. It is an excellent lubricant especially for airplane engines. When hydrated it is converted into a quick-drying oil used in paints and varnishes. Caster oil may also be used in making soap, inks and plastics, for preserving leather and as an illuminant. The leaves have insecticidal properties, while the stalks are a source of paper pulp and cellulose. The oil cake or pomace is poisonous but makes and excellent fertilizer.

Other Nondrying Oils

Of minor importance are kapok oil from the seeds of the kapok tree and used as a substitute for cottonseed oil; tea-seed oil from *Camelia sasanqua*, a valuable oil in China, Japan and Assam; and oil of ben from *Moringa oleifera*. Nondrying oils are also obtained from almonds, pecans, filberts, pistachio, and pili nuts and from the flesh of avocados.

Vegetable Fats

Coconut Oil

A widely used fatty oil, coconut oil is obtained from the dried white interior of the seed of *Cocos nucifera*. This oil is pail yellow or colourless and is solid below 74 deg. Fahrenheit. Following harvest, the nut husks are removed and the interior split open and dried either by natural or fire heat. The dried tissue, or copra, is then easily removed. This is ground up and the oil expressed. The cake is sometimes placed in a hydraulic press a second time, and still more oil is removed. The yield is 65-70 percent. Fresh coconut meat can be pressed also giving a yield of 80 percent or more oil. Refined coconut oil is edible and is now extensively used as a food product. It is especially well adapted for the purpose of margarines because it is solid at room temperature. It is used in candy industry and for making the highest quality soaps, cosmetics, salves shaving creams, shampoos, suntan lotions, etc. It is used for marine soaps and also as an illuminant. The cake is an excellent livestock feed. Copra is produced in Ceylon, India, Polynesia, the

Philippines and the West Indies. Most of the oil is expressed in Europe, the United States and Japan, though Ceylon and India export large amounts also. In the latter part of the 20th Century there has been a trend away from consuming coconut fat because of possible harmful effects from saturated fats..

Palm Oil

This is a white vegetable fat that is solid at room temperature and which is obtained from the seeds of the African oil palm, *Elaeis guineensi*, a tree of Western Africa. It has been transported worldwide and is cultivated in Haiti, Brazil and Honduras, where it is called dende. Oil palms are very productive, beginning to bear at the age of 5-6 years and reaching full bearing at 15. They continue to bear until 60-70 years of age. Each tree has 10 bunches of 200 nuts a year. The fibrous pulp of these fruit contains 30-70 percent fat.

The oil was obtained mainly by crude methods for years until newer methods were developed. It is yellow-orange or brownish red in colour. Most of the supply has been from Sumatra, Java and Western Africa. Palm oil is used in making soap and in the manufacture of tin plate, terne plate and cold reduced sheet steel.

The refined oil is used in margarine and vegetable shortenings. It has also bee used as a fuel for diesel motors in Africa. Health conscious people have refrained in consuming this fat because, like coconut, there may be harmful effects from saturated fats.

Palm-kernel Oil

This white and valuable oil is obtained from the kernel of the African oil palm, *Elaeis guineensis*. It was extensively used in the margarine and candy industries because of its pleasant odor and nutty flavour. It has also been used for making blycerin, soap, shampoos and candles. Natives in West Africa express some oil for their own use, but most of the kernels are shipped overseas. Palm-kernel cake is a good cattle feed.

Brazilian Palm Oils

During World War II when supplies of copra and palm oil were reduced, attention was directed to the rich sources in Brazil. This included over 500 species of palms, many of which are potential sources of oil. Although not native to Brazil, both coconut and the African oil palm occur there in great abundance. Additionally, several native species of palm have economic importance. These include the babasu, cohune, lieuri, tucum and murumuru palms, all of which provide kernel oils.

Babassu Oil

The most important of the New World palm oils is obtained from the babassu palms, *Orbignya martiana* and *O. oleifera.* These are magnificent trees 60 ft. in height with a vase-shaped crown of leaves. Two to eight enormous clusters of fruit are produced, each weighing about 200 lbs. and containing from 200-600 fruits. The outer portion of the fruit is a dense, tough fibrous husk. This surrounds a thin mealy mesocarp and the nut, which has a thick very hard shell. This industry was retarded when only hand labour was available to crack the nuts, but machines were developed that can exert 10,000-25,000 lb. pressure.

The nuts contain from 2-6 kernels with a 63-70 percent oil content. Babassu oil is expressed from these kernels and when refined is used as a substitute for coconut oil. It is also used in making bulletproof glass, lubricants, and explosives and as a fuel for Diesel engines.

Cohune Oil

Nuts of the cohune palm, *Orbignya cohune*, native to Central and South America, contain 40 percent of a firm yellow fat. At one time there were over two million acres of this palm managed in British Honduras that yielded from 1,000-2,000 nutts per tree. Like the babassu, the nut is very hard to crack, but machinery has been developed.

Licuri Oil

The licuri or ouricuri palm, *Syagrus coronata*, occurs in dry areas of eastern Brazil. It is important commercially as the source of a palm-kernel oil and also of a wax that occurs on the leaves. The fruits resemble miniature coconuts.

Murumuru Oil

The murumuru palm, *Astrocaryum murumuru*, is the main source of palm-kernel oil in the sate of Para, Brazil. The tucum palms, *Astrocaryum tucuma* and *A. vulgare*, or northern South America yield both a palm kernel and a pulp oil and also a fibre of commercial value.

Vegetable Fats of Minor Importance

Cocoa Butter

This is the white or yellowish fat with a chocolate odor and flavour that is expressed from the beans of the cacao or cocoa, *Theobroma cacao*, during the process of making cocoa. It is firm at room temperature. It is used mainly for making chocolate but also for cosmetics and in perfumery and medicine.

Carapa Fat

This is a thick white or yellow oil obtained from the seeds of several species of the genus *Carapa* and is used for soap and as an illuminant. Natives in South America have used the oil from *C. guianensis* to grease their bodies and drive off insects. *Carapa moluccensis* is a native of East Africa, India, Sri Lanka and the Moluccas

Shea Butter

Seeds of *Butyrospermum parkii* from Africa provide shea butter. It is a greenish-yellow fat with a pleasant odor and taste. The fat is also used mixed with, or as a substitute for, cocoa butter in chocolate manufacture. Inferior grades are used for soap and candles. A large amount was exported to Europe in the 20th Century.

Mowra Fat

Two species of the genus *Madhuca, M. indica* and *M. longifolia,* are the source of various products in India. They are best known as mowra fat, bassia fat, mahua butter of illipe butter. The trees are both wild and cultivated. The kernels contain 55-65 percent of soft yellow oil that is widely used locally for cooking and tallow. Over 66 million pounds were exported to Europe for use as a margarine and chocolate fat and in soap manufacturing during the 1950's. The cake is not fit for food but is used as a fertilizer. *Madhuca butyracea* is the source of a similar product, phulwara butter, also used locally.

Borneo Tallow

This is a hard yellowish-green and brittle solid fat sometimes known as green butter. It is obtained from *Shorea aptera* and some other species of the same or allied genera native to the East Indies. The kernels contain 50-70 percent fat. They are dried and expressed by the local people for their own use or are exported to Europe for soap making and as a substitute for cocoa butter.

Chinese Vegetable Tallow

This occurs as a thick layer of hard white fat on the seeds of *Sapium sebiferum* a tree from China. It has been introduced worldwide including the southern United States. After treatment the tallow is used in soap and candle making. The seeds contain a drying oil that has some value.

Nutmeg Butter

Seeds of nutmeg, *Myristica fragrans*, and related species contain about 40 percent of a yellow fat with the flavour and consistency of tallow. Nutmegs hate are not suitable for use in the spice trade are roasted and powdered and the oil extracted between warm plates. Several varieties of nutmeg butter have appeared in the market, all used for ointments or for candles. Mace yields a similar material.

Other Vegetable Fats

There are many other vegetable fats used locally. Among these pongam oil from the seeds of *Pongamia pinnata* is used for illumination and medicine in India and Sri Lanka; macassar oil from the seeds of *Schleichera oleosa*, a soft yellowish-white fat in India, Sri Lanka and the East Indies for coking purposes, as a hair oil and for illumination; and ucuhuba and otoba butter obtained from species of *Virola* in northern South America.

Members of the Flacourtiaceae in Asia, Afric and South America have oily seeds that contain chaulmoogric and usually hydnocarpic acid. The natives in treating skin diseases have used these oils. The most important one, chaulmoogra oil, are discussed under Medicinal Plants

Waxes

Waxes usually occur on the epidermis of leaves and fruits where they serve to prevent water loss through transpiration. The impervious waxes are harder than fats and have a higher melting point. They do not become rancid and are less easily hydrolyzed. Chemically waxes are similar to fats, but they are esters of monohydric alcohols rather than glycerides. Very few are of commercial importance.

Carnauba Wax

Carnauba is the most important vegetable wax. It occurs as exudations on the leaves of the wax palm, *Copernica cerifera*, native to Brazil and other areas in tropical South America. The slender fan palm is called "Tree of Life" locally because almost all parts of the pant are useful. The commercial supply of carnauba wax is obtained from wild trees in northeastern Brazil. Young leaves are carefully gathered before they are fully open. They are dried in the sun for several days until the wax has turned to a flourlike dust. This is then removed by threshing. It is then melted down in clay vessels, followed by straining, cooling and forming into cakes or broken into small pieces for shipment. There are several grades. The crude product is greenish-gray and very hard with a high melting point.

Carnauba wax is used in the manufacture of candles, soap, high-luster varnishes, paints, carbon paper, batteries, sound films, insulation, salves, ointments and previously phonograph records.

The trunk of *Ceroxylon andicola* of the Andean Region produces a similar wax, which has been used as a substitute.

Candelilla Wax

Candelilla is from *Euphorbia antisyphilitica*, a low, light blue-green desert shrub of Texas, Mexico and Northern Central America. The tiny leaves are quickly deciduous. The wax exudes from pores and forms a thin film on the stems. The amount produced increases during winter, so that the plants are collected in that season. The principal source is from wild plants. The wax is extracted by solvents or by boiling. The crude material is white, but the refined product is a light tan colour and has a sweetish odor. Candelilla wax is softer, contains more resin and has a lower melting point than carnauba, and is therefore less valuable. It has been used as an extender wax in a mixture with others. Chemical treatment improves its quality, however.

Myrtle Wax

Berries of the bayberry, *Myrica pensylvanica*, and the wax myrtle, M. cerifera, are both native to the eastern United States, and they are covered with a thick layer of wax. The material is in reality a fat rather than a wax. It is removed from the fruits by boiling in water and is used for candle and soap manufacture.

Cauassu Wax

The leaves of the Cauassu, *Calathea lutea*, a tall herb of the lower Amazon region, are potentially an important source of a commercial wax. The wax occurs on the underside of the large leaves. Drying in the sun for only 2-3 hours is enough for the very thin scales to form, which are then removed by scraping. New wax-yielding leaves are produced within a year, while it takes 8-10 years for carnauba leaves to be renewed. Cauassu wax is similar to carnauba and can be used for the same purposes.

Jojoba Wax

Jojoba is an evergreen bush, *Simmondsia chinensis*, of semiarid regions on the southwestern United States and northwestern Mexico. it is unique in having seeds with a 50 percent liquid wax content. This wax is suitable for polishes, candles and as a substitute for whale oil. It is also useful as an illuminant and, after processing, for several types of foods. The oil cake is an excellent livestock feed and the acornlike fruits are edible. Cultivation of jojoba increased dramatically in the United States during the latter part of the 20th Century, but has declined thereafter.

Other Waxes

Berries of the wax tree of Japan, *Rhus succedanea*, and the leaves of raphia and licuri palms, sugar cane and esparto may also furnish commercial wax.

Soap Substitutes

Many plants contain natural products that can serve as substitutes for soap. These are the saponins, a group of water-soluble glucosides. Such plants yield soap froth in water that forms emulsions with fats and oils and are capable of absorbing large amounts of gases such as carbon dioxide. Besides the few that are commercial important, there are many wild species that have been used locally. The most important saponin-containing plants are the following:

Soapbark

Quillaja saponaria, the soapbark tree, grows on the western slopes of the Andes from Peru to Chile. The commercial material is derived from dried inner bark, which is removed after the outer bark has been shaved off. The saponin content of the bark is about 9 percent. Soapbark forms a copious lather in water and is used in washing delicate fabrics. It was one of the best emergency materials for cleaning lenses and precision instruments during World War II. In medicine it has been used as an expectorant and emulsifying agent. However, it is a dangerous drug to take internally. Being very toxic it tends to dissolve the blood corpuscles. Therefore, its use to increase the foaming power of beer and other beverages is discouraged. Soapbark is also a good coetaneous stimulant and is often used in hair tonics.

5

Forest Products: Wood and Cork

Forest products have been of service to humans from the very beginnings of our history (Hill 1952). The most familiar, and the most important, of these products is wood. Wood is used in all types of construction, as a fuel and as a raw material of the paper and rayon industries. Other products include rubber, cork, many of our tanning materials and dyes, resins gums, oils, drugs and even sugar, starch and some chemicals.

Additionally, the seeds and fruits of many trees may serve as food for humans or their livestock. In addition to being of value to humans, forests themselves have many utilitarian features. They help to regulate climate and temperature. They aid in the conservation of the water supply and in flood control by preventing water runoff.

Their roots hold the soil firmly in place and control erosion. They may also act as shelter area against drying winds. They afford a range for livestock, a shelter for wild life and offer many recreational aspects for humans, the importance of which cannot be underestimated. This section is limited in discussion to wood and cork following the format of Hill (1952). Other useful materials from trees are considered in other sections together with similar economic products from other sources.

Wood Structure

Wood is a plant secondary tissue that is formed mainly in the stems of gymnosperms and dicotyledons through the activity of a growing layer, the Cambium. The cambium is responsible for the growth of stems in thickness through the formation annually of new layers of both wood and bark.

Composition of Wood

Wood is a heterogeneous tissue made up of several different kinds of cells, some that have the function of mechanical support and others that of conduction. In softwoods, or gymnosperms, both these functions usually occur in cells called Tracheids. In hardwoods, or angiosperms, a division of labour usually exists. Mechanical support is provided by the several types of wood fibres that make up a greater part of the woody tissue, while the conduction of water is by tubular cell fusions called Vessels. Tracheids are also occasionally present. Wood also functions in the distribution and storage of carbohydrate that is accomplished in the Parenchyma Cells. These are the only parts of the wood that is truly living and which contains protoplasm. Parenchyma cells occur in two forms, (1) wood parenchyma and (2) ray parenchyma. The former are arranged vertically in the stem while the latter are horizontal.

Diagnostic Features

Valuable diagnostic traits are afforded by the arrangement of the different types of cells in wood. Woods may be distinguished by pores, early wood and late wood, growth rings, rays, heartwood, sapwood grain and figure.

Porous & Nonporous Woods

The presence or absence and the nature and arrangement in cross section of pores that are really vessels, offer a quick way to classify woods. Conifers that do not have vessels are in the nonporous wood category. Hardwoods that do have vessels may be further divided into those where the pores are arranged in concentric circles, the outer and inner portions of which differ according to number and size of pores, and those where the pores are all small and about the same size and are scattered uniformly through the wood. The first category is called ring-porous and the second diffuse-porous. .

Early & Late Wood

In temperate climates new wood is formed annually during a limited growing season and definite growth layers result. These usually have two distinct areas within each layer. In springtime when growth resumes, the first wood to be formed contains many large and thin-walled cells as a response to the greater need for conducting nutrients.

This is the early wood or spring wood. As the season progresses a more dense kind of wood is laid down that has smaller, thicker walled cells, the late wood or summer wood. This produces a sharp transition between the cells formed at the end of a growing season and those formed at the beginning of the succeeding one. In cross section this appears as concentric rings called Growth Rings. The growth ring of one year is called an annual ring and the number of these indicated the tree's age. In the tropics where growth may continue throughout the year growth zones may occur also, but they are due to changes in weather or other causes rather than to definite growth periods.

Rays

These are thin sheers or ribbons that are made up primarily of parenchyma cells oriented at right angles to the stem's main axis. They vary in height, width and arrangement. In cross section they are visible as lines that radiate from the stem's centre. They are most obvious in radial sections where they can also be used to identify the tree species by the variety in their form and arrangement.

Sapwood & Heartwood

When young all wood cells are physiologically active. But in time many of them lose their activity and become skeletons that serve only to provide strength to the tree. Eventually two distinct areas develop: (1) a light coloured outer region of varying width, the sapwood, and (2) a darker inner region the heartwood. Only parenchyma cells in the sapwood remain physiologically active. The older cells of the heartwood that have died often attain a colour and are very resistant to decay because of the deposition in them of gums, resins or other waste substances. This heartwood can be polished to a high luster and is valuable in making furniture, cabinets and other woodworking aspects. Although heartwood is generally distinct from sapwood in durability, appearance, etc., it may not always be very clearly differentiated.

Grain & Figure

These along with texture are terms that are often used indistinctly. Texture usually refers to the relative size and quality of the various parts of wood, while grain refers to their structural arrangement. Figure applies to the design or pattern that appears on the surface of lumber and may be caused by the kind of grain, the presence of colouring material that have penetrated the tissues or both.

There are many different kinds of grains. In straight-grained wood the various parts occur parallel to the stem's main axis. When they are twisted spirally about the axis they make up spiral-grained wood. When the longitudinal course of the different parts undulates slightly, a wavy grain is the result. Curly grain is due to various growth irregularities and other causes.

The distinctive figures that wood often shows and that make it valuable for decorative purposes are due mainly to the different types of grain in combination with the rays, rings, sapwood, heartwood and many other arrangements of cells. By cutting the wood in different ways these variations may be stressed in one way or another. In quartersawing wood is cut parallel to the rays and across the rings, while in plain sawing the wood is cut at right angles to rays and tangent to the rings. Sometimes figures in the wood are caused by masses of colouring material that has penetrated the tissues and which may occur in zones or streaks. Snakewood, for example, has streaks on the tangential surface suggestive of the markings on a snake's skin.

Wood Mechanical Properties

Wood has some mechanical characteristics that either alone or in combination determine its utility and fitness for various purposes. These characteristics differ among species and even in individual trees. The mechanical properties of wood are those that enable it to resist various external forces that would changes its shape and size and produce deformities. Such external forces induce internal resistance, called stresses, in the wood. When they exceed the force of cohesion among wood elements, a failure will occur. Some of the more important mechanical properties are toughness, strength, stiffness, cleavability and hardness.

Strength

This designation ought to be restricted to the ability to resist certain forces, such as pulling, crushing and shearing. In addition the word should always be modified to indicate the specific type of resistance involved.

Crushing Strength

Also called compression strength this is the resistance offered to forces that tend to crush wood. Such forces may be applied endwise, and so parallel to the grain, as in a column; or sidewise, where they are at right angles to the grain, as in the case of railroad ties. The highest crushing strength is endwise.

Tensile Strength

This is the resistance to forces that would cause the wood to be pulled apart. They also may be applied either parallel to or at right angles to the grain. The maximum tensile strength exists when the force is parallel to the grain and it is 2-4 times grater than the crushing strength.

Shearing Strength

This is resistance to forces that tend to make the fibres slide past one another. Such forces may be applied parallel to the grain, at right angles to it or obliquely. Wood is most resistant to a perpendicular shear.

Cross-breaking Strength

Also known as bending strength it is usually applied to beams or other parts of timber that are supported at both ends and loaded between these points. The strength involved is the resistance to forces that cause the beam to break, and all the other forces mentioned above are also involved.

The upper part of the beam is under compression; the lower under tension, and shearing also is involved. Because of a greater tensile strength, a beam fails first by compression, or buckling, on the upper side. As the load increases, the tensile strength gradually fails until the beam snaps across on the underside. Wood strength is the most important factor in deciding the value of any particular type for structural purposes. This varies greatly, however, and is influenced by the wood density, the moisture content, the presence of defects and other factors.

The relationship between density and strength is especially close and density is considered to be the best criterion of strength. Among the strongest woods in North America are White oak, longleaf pine, larch, hickory and sugar maple.

Stiffness

This is the measure of the ability of wood to resist forces tending to change its shape. It is the capacity to withstand deformation under a bending strain. It is often contrasted with flexibility that is the ability to bend without breaking and it involves toughness and pliability.

Toughness

This is a combination of other properties that is usually considered to mean the ability of wood to absorb a large amount of energy and to resist repeated, sudden, sharp blows or shocks. A tough wood is difficult to split and although it might rupture it does not break easily.

Hardness

The measure of the power of wood to resist indentations, abrasions and wear. It is variable with density and determines the ease with which wood can be cut or sawed.

Cleavability

An expression of the ease with which wood may be split. This is desirable for firewood and undesirable when wood must be able to hold nails or screws. Wood tends to split more easily along the rays and when its grains are straight.

Factors Influencing Mechanical Properties

There are many factors involved in the strength, hardness and other properties of wood, with some of the more important ones being density, moisture and some defects.

Density

Density differences are among the main causes of variations in the mechanical properties of wood. Because of the close relationship between the two properties, the density of any wood may be the best indication of its strength. Density is usually expressed as specific gravity or the relative density. Density is simply the weight of a unit volume of any substance that is expressed as weight per cubic food; but specific gravity is the ratio between the weight of the substances involved and that of an equal amount of pure water.

The density of wood substance is almost the same in all species and its specific gravity has been estimated at around 1.55. This shows that wood is 1.55 times more heavy than water and is explains why a piece of wood will sink as soon as the air in its cavities has been replaced by water. However, the vasriations in density that are observed in different species, individuals and even parts of the same individual, are due to differences in the actual amount of wood substance present. It is the amount of cell wall material compared to cell cavities. Woods with thicker walls and smaller cavities are denser and heavier, while the more porous woods with larger cavities and thinner cell walls are lighter and less dense.

Specific gravity of wood can be determined in different ways, but usually the weight is calculated from oven-dry material and the volume is measured with the wood in any desired condition. Nevertheless, because the moisture content of green, partly seasoned, and seasoned wood is different, and since this difference affects the specific gravity, it is obvious that specific gravity values mean little unless the condition of the wood is clarified.

Because the amount of actual wood substance has an important bearing on the density of a piece of wood, it can be assumed that heavier woods are therefore stronger. This is not always the case because the presence of resins, gums and other infiltrated substances and also the amount of water may affect the weight without altering the strength.

Hickory, Osage Orange, Oak and Persimmon are some of the heaviest native woods. Tropical woods show great diversity in weight as exampled by lignum vitae being very heavy and balsa being very light.

Moisture

Wood contains variable amounts of water that differs according to species and environmental conditions from 40-100 percent of the dry weight. The water occurs in wood cell cavities or in the cell walls, where it is known as hygroscopic water. The amount of hygroscopic water required to saturate the walls is fibre-saturation point and constitutes 20-35 percent of the dry weight. The variation in the amount of water that is present in wood is due to a number of things and is made possible by a typical property inherent in the wood. This property, or hygroscopicity, is the ability to absorb or give off water under different conditions, with an accompanying swelling or shrinking.

Moisture content of wood influences its weight, density and frequently its strength. If the amount of water present is above the fibre-saturation point, the weight is increased but the strength is not changed. However, if the amount of water is brought below the fibre saturation point through evaporation, then the strength and other mechanical properties as well are changed. This means that it is only the loss of hygroscopic water that is responsible for the increase in strength that accompanies seasoning, as the drying out of wood is called. This loss of hygroscopic water causes the wood to shrink, due to changes that take place in the cells.

As the water leaves the walls contract, the cells become more closely compacted, and the fibres become stronger and stiffer. This tendency of wood to shrink as it dries is one of the huge drawbacks to its use. The amount of shrinkage varies under different conditions and is likely to occur unevenly. Wood shrinks very little lengthwise, and only about half as much radially as tangentially.

This result of uneven shrinkage can contribute to warping, checks, shakes and other defects. These then counteract any increase in strength. Despite such shortcomings, dried or seasoned wood is usually stronger, harder, stiffer and more durable than unseasoned wood. Artificial methods of seasoning are deployed in order to control the process.

Artificial Seasoning

Two principal types of artificial seasoning are kiln drying and air seasoning. In air seasoning, the moisture is removed by exposure to air without resorting to artificial heat. It is done in the open until the wood ceases to lose weight. The final moisture content varies 12-30 percent with the species, the duration of the process and environmental conditions. The main objects of air seasoning are to reduce the weight, the amount of shrinkage and possible defects; to render the wood less subject to decay; to increase its strength and combustibility; and to prepare it for painting, preservative treatment and kiln drying.

Kiln drying heat is applied to wood in an enclosed space. Either seasoned or unseasoned wood can be used. The moisture is removed more rapidly and completely, the moisture content of the finished product varies from 4-12 percent. Kiln drying applied to green lumber often prevents checks, warping and defects due to fungi or insects.

Defects

Mechanical properties of wood are often affected by the presence of various types of defects that may be of some importance. These defects are due to many causes. Some of them may be normal characteristics, but ones that limit the usefulness of wood. Wood is dimensionally unstable; it swells, warps, and checks with humidity and temperature changes. Its strength is unidirectional in that it is strong with the grain and weak across the grain. Such disadvantages may be surmounted partially by the use of plywood, wood alloys, or reconstructed wood.

Certain defects may develop during the seasoning of wood. Defects such as knots and cross grain, may be inherent in the wood structure and others may be due to external causes. Among the latter are fungi, insects, marine borers, parasitic seed plants, birds, lightning, frost and fire. Defects caused by insects and fungi are of greatest concern.

Insect Damage

Insect damage may be much greater than generally realized. All sorts of wood from standing timber to lumber and wood products may be attacked. Holes produced by wood-boring insects make up the principal type of injury. Insects are most destructive as larvae. By the end of the 20th Century, the ravages of termites have been increasingly serious.

Decay

Fungi cause wood to decay. Four conditions are necessary for the development of these lower plants. Unless a favourable temperature, sufficient moisture, at least a small amount of oxygen and an adequate food supply are available, fungus decay cannot occur. The food is furnished by cellulose and lignin in cell walls, and is made available by enzymes that are secreted by the fungi. Brown rots remove the cellulose, leaving behind a brittle brown mass of lignin compounds; the white rots utilize the lignin and leave the white cellulose behind. Other fungus species are able to utilize both cellulose and lignin.

Woods vary in their natural resistance to fungi. This property is known as durability and is so important that it alone may determine the ultimate use of wood, particularly in the case of poles, posts and mine timbers that are exposed to moisture. Sapwood decays quicker than heartwood because the latter usually contains resins, gums, tannins and other substances that resist fungi. Seasoned wood is also less apt to decay because of the lower moisture content. Among native woods that are naturally very durable are the redwood, cedars, cypress, locust and osage orange. The least durable woods include balsam fir and basswood (linden).

Frequently wood is rendered more immune to decay by treating it with preservatives that are poisonous to fungi. This process of wood preservation has developed into a considerable industry. Various chemicals are used as preservatives, mainly creosote and zinc chloride. The methods used are brushing or spraying the surface, dipping in open tanks and various pressure processes that make possible a deeper penetration of the preservatives.

Wood Uses

Wood is used in so many ways that it is impossible to discuss all in detail. In many countries there is a great deal more wood cut each year than is replaced by normal growth, which does not bode well for a continuous supply in the future.

Forest Products as Fuel

Fuel is an indispensable necessity of life both in home and industry. Any material that burns readily in air can be utilized, but this includes a great variety of plant products. The most important of these are wood, peat and coal, which represent different stages in the carbonization of the original plant tissue.

Wood

Farms and rural communities have accounted for about 90 percent of the total amount of wood used for fuel. Wood makes an excellent fuel because it is about 99 percent combustible when dry and so leaves only a small amount of ash. It is also flaming fuel and well adapted for heating large surfaces. The value of different kinds of wood for heating purposes depends on the amount of moisture present. Therefore, seasoned wood is better than green wood. Hardwoods have the greatest fuel value, particularly such woods as hickory, eucalyptus, oak, beech, birch, maple and ash. Longleaf pine in the southern United States is mainly used while in the western area Douglas fir, western yellow pine, western hemlock and western larch are used.

Peat

Peat is made up of deposits of vegetable matter that have accumulated in swamps and bogs and slowly decomposed, becoming somewhat carbonized and compacted. The various plant tissues can still be discerned. The process of peat formation is continuous, and peat is a valuable fuel in countries where wood is scarce. It is more bulky to manipulate and leaves from 5-15 times as much ash. At the lower depths of some peat bogs a soft brown coal called lignite may be found. This also has the original plant structures still visible.

Coal

Coal comprises the fossilized remains of plants that lived in former geological periods (Hill 1952). The original plant tissue has been more fully decomposed and converted into carbon. Coal is much harder and more compact than peat or lignite, and has a greater heating power. It also yields a larger amount of smoke and ash. Anthracite or hard coals are the oldest and contain about 95 percent carbon. Bituminous or soft coals are more recent in origin and thus are less completely carbonized. They tend to soften and fuse at temperatures below the combustion point. Cannel coal consists of fossilized spores. It is very compact and oily and burns with a candle like flame. Unlike other coals it does not soil in one's hands.

Coal is a comparatively inexpensive source of power and heat and also of many useful chemical products. Among the latter, which are obtained by destructive distillation, are oils, such as benzol and naphtha; coal gas that is used for fuel and illuminating purposes; ammonia; coal tar, the source of dyes, antiseptics and many other materials; and coke

Coke bears the same relationship to coal that charcoal does to wood. It is obtained by the smothered combustion of coal in piles or special ovens, usually as a by-product of the illuminating-gas industry. It is almost pure carbon and burns without smoke or flame. Coke is an excellent fuel that is especially used in metallurgy.

Petroleum

There are no traces of the original structure present in petroleum, and it has been generally believed that petroleum had an organic origin and was formed under pressure from the minute floating plant and animal life of former shallow seas. Crude petroleum has many uses, but the substances derived from it by fractional distillation are of much greater importance. Among these products are gasoline, kerosene, plastics, petroleum jelly, medicines and paraffin.

Lumber

Lumber from wood has been in use for building purposes and other construction since early times. In the United States the first sawmill was established in Maine in 1631. Thereafter a huge industry has developed. The word "lumber" refers to wood that has been prepared to some extent for future use. The larger pieces of lumber that are used in heavy construction are often called "timber." The standard unit of measuring lumber is the board foot, which is the equivalent of a piece of wood 1 in. thick, 12 in. wide and 1 ft. long.

Lumber Industry

The location of the lumber industry and species utilized are usually in constant change. In the United States it has always been in a region where large stands of virgin timber were available. Until 1830 the state of Maine was the main lumber-producing area, and for the next 40 years New York and Pennsylvania took the lead. By 1870 the centre shifted to the Lake States, with first Michigan and later Wisconsin.

After 1910 the Southern states became the leading producers, with southern pine replacing the northern hardwoods and white pine. Then the centre moved to the Pacific Northwest and began utilizing the immense stands of Douglas fir and other softwoods. The Southern states continue to produce most lumber, although Washington and Oregon are still prime providers.

There have been over 150 native American species utilized in the lumber industry. However, the softwoods have furnished about 78 percent of all the lumber cut. For many years the eastern white pine, *Pinus strobes*, was the outstanding timber tree and this was one of the most valuable trees in the world. The demand for white pine was so great that the supply soon became diminished and was eventually replaced by other species. Oak and hemlock have also had a prominent role. By the end of the 20th Century the most important tree species were southern yellow pine, Douglas fir, western yellow pine, oak hemlock, white pine and red gum. Douglas fir and yellow pine produce twice as much as all the other combined. Many woods that at one time were considered worthless have later become important. Examples include sycamore, beech, red gum and tupelo.

Lumber Utilization

Usually about eight percent of the lumber cut each year is exported; 32 percent is used for structural timbers and rough lumber for construction; 33 percent goes to the planning mills and 27 percent is used in other woodworking industries.

Export Lumber

Lumber has been an important North American foreign trade since the early days of European colonization The demand for American timber has been especially great in Japan.

Structural Timbers

These are the large sizes of lumber from sawmills that are used for buildings, bridges and other types of heavy construction. They include girders, stringers, beams, joists, rafters, posts, caps, planks, caps, roofing, boards for sheathing, and flooring. In the early days before steel, immense quantities of heavy timbers were used in ship building. Structural timbers are obtained mainly from the softwoods because large sizes are readily available. The timbers are usually sawed from the heart of the tree, and even though defects may be present, the bulk is so massive that the strength is not impaired. Strength is the main requirement of a good structural timber, especially the resistance to stresses that can be estimated. Numerous timber testing experiments have made possible the estimation of the working stresses with great accuracy. Durability, soundness and ease of working are other desirable qualities. The main species used in the United States for structural timbers are southern yellow pine, eastern white pine, hemlock, Douglas fir, western yellow pine, spruce, redwood and larch.

Planing-mill Products

These are usually associated with sawmills and they use a large amount of lumber representing over 60 different species. Their products are sometimes classified as "factory lumber," that is lumber which has been recut to smaller dimensions and reworked. The principal products of the planning mills are doors, sashes, window frames, blinds and interior finishes. The best wood for millwork has a straight grain and a soft uniform texture. It should not shrink or swell and it should be easy to work and capable of taking varnishes and paints. The less expensive products are made from white pine, Douglas fir, yellow pine and other softwoods. Expensive items include birch, oak, red gum, maple, walnut, mahogany and other hardwoods with a fine figure. Veneers are now being used extensively for door panels.

Interior finishes include baseboards, columns, cornices, mantels, grills, stair work, posts, balusters, scrollwork, porch work and trimming. Most of the North American and imported woods that are decorative and have good wearing qualities are used for these purposes.

A more recent development of the planning mills has been hardwood flooring. Only the most durable woods that also have attractive designs are used. These are kiln dried and seasoned. The woods used include beech, maple, oak, tupelo, birch and yellow pine.

Other Woodworking Industries

There are numerous industries that have many variable requirements and products. Most use only a relatively small amount of wood, but a few of the more important industries include railroad car, box and crate, furniture, vehicle, agricultural implement and woodenware industries.

Boxes & Crates

Items such as crates, baskets, boxes and other containers are used for the transport of canned goods, farm products and many other articles. For box-making the wood should be light, east to work, strong, with good nail-holding power and a surface that can be printed upon. Lower grades of softwood lumber and the softer hardwoods are mainly used. The principal species are southern pine, western yellow pine, red gum, hemlock, white pine and spruce.

Furniture & Fixtures

This requires wood of special hardness, strength and durability. It should not shrink or warp, and must be ornamental and capable of a high polish. Birch, Maple, Oak, Red Gum, Walnut are the principal species, although chestnut, beech, Elm, tulip and basswood are also used. A large number of imported woods that have attractive colour or figure are utilized also. There have been over 60 species used. Veneers are of increasing importance for they can be used to cover less expensive and less attractive woods.

Railroad Cars

There has been a lot of lumber used annually for the construction of railroad cars with over 40 different species of wood being used. Oak, Maple, Hemlock, Cypress and several ornamental species are of some importance although Yellow Pine and Douglass Fir have been used the most. By the 21st Century railroad cars have begun to be constructed with plastic walls and that has greatly diminished the need for wood in this industry.

Vehicles

Wood was used extensively in the manufacture of vehicles in the first half of the 20th Century, but has been largely substituted by plastic in the 21st Century.

Misc. Industries

Other industries using wood during the industrial age have been agricultural implements, caskets, coffins, refrigerators, kitchen cabinets, ship and boat building, matches, woodenware and novelties, musical instruments, tanks and silos, signs and supplies, professional and scientific instruments, electrical machinery and apparatus, machine construction, toys, laundry appliances, handles, supplies for dairymen, poultrymen and beekeepers, tobacco boxes, patterns and flasks, sporting and athletic goods, boot and shoe findings, shade and map rollers, brooms and carpet sweepers, picture frames and moldings, motion-picture and theatrical scenery, brushes, plumber's woodwork, shuttles, spools and bobbins, trunks and valises, sewing machines, pumps, wood pipe and conduits, airplanes, toothpicks, printing materials, playground equipment, dowels, clocks, paving materials, saddles and harness, gates and fencing, butcher blocks and skewers, bungs and faucets, firearms, scales, elevators, whips, canes and umbrellas, tobacco pipes and artificial limbs. By the 21st Century either metal or plastics have substituted most of these.

Posts

Fence posts are used mainly on farms and along roads and railroad right of ways. The old rail fences have almost disappeared. Posts are usually cut 7 feet in length and from 4-6 inches in thickness. They are used in the round or are split. Strength, light weight and durability in the soil and weather are the main requirements. Woods that have been utilized for posts are mainly cedar, redwood, chestnut, oak, tamarack, black locust, ash, osage orange and cypress. Posts are usually treated with preservatives before being placed into the ground.

Mine Timbers

There is much wood used in mines for shafts and for supporting structures, including collars, caps and props. Of particular importance for safety are strength and durability, and the wood is used in an environment that is conducive to decay. Woods that are normally used for this purpose are pine, oak, tamarack, chestnut, beech hemlock, maple and Douglas fir. These timbers rarely enter the trade for most of the supply is obtained locally from whatever is available.

Poles & Piling

Many poles are used annually to suspend electricity, telegraph and power-transmission lines. Similar amounts are used for work in harbours including trestle and bridge construction. These poles are used in the round and because they are prone to decay at ground level, only those species with durable sapwood are preferred. These are also treated with preservatives. Strength, light weight and accessibility are prime qualities, and shape is also important. In addition to these stated requirements, pilings must be able to withstand heavy top loads and can withstand blows as they are pounded into the ground. Species preferred are lodgepole pine, Douglas fir, southern pines and cedar. Sometimes chestnut, cypress, oak, larch, redwood, locust and elm are also used.

Cooperage

The manufacture of wooden containers bound together with hoops of wood is an old industry that dates to Biblical and Roman times. However, production has been in a steady decline through the 20th Century due to the competition from other types of containers. Two main classes of cooperate are slack cooperage, to hold dry substances, and tight cooperage, to hold liquids.

Slack Cooperage

This includes a great variety of barrels, casks, tubs, pails, buckets, churns, kegs and other containers. They have been used to transport meat, fish, tobacco, fruit, flour, vegetables, cement, sugar, glassware, crockery and etc. A single barrel usually consists of 15 staves, one set of heading and six hoops.

These various parts may be manufactured in different factories and regions. There are many grades of slack cooperage that ranges from sugar and flour barrels, with tongued and grooved staves, to the loose-fitting and less expensive cement barrels.

Wood to make slack cooperage should be inexpensive, easy to work, light, elastic and free from twisting and warping. Limbs, tops, defective logs and other forms of waste lumber may be used. Veneers have been used to some extent. Heads and staves are usually made from pine, red gum, beech, maple, oak, Douglas Fir and ash.

Tight Cooperage

Kegs and barrels to be used for containers for beer, oil, wine and other liquids require more careful construction. Woods that will impart no taste or odor to the liquids and which are impermeable are essential. White oak has been the main wood used, especially when the liquids are to remain the barrels for a long time. Red oak, red gum, yellow birch, white ash, sugar maple and Douglas fir are frequently used. The finished product is then usually treated on the inside with paraffin to insure that no leakage will occur. The hoops are mainly of strap steel.

Heavy Cooperage

Vats and large tanks constitute heavy cooperage. They are made with staves and heads of white oak, cypress, Douglas fir or redwood and bound together with metal straps.

Railroad Ties

The average life of an untreated railroad tie is only 5-6 years, so that the demand continues for replacement ties. These are usually hewn from a seasoned wood but they may be sawed. Strength, durability in the soil and the ability to resist impact, crushing and spike pulling are important traits. Treated ties are just as serviceable as those made from naturally durable woods. Most ties in North America are made from southern pine, red gum, oak and Douglas fir. Other species include cedar, chestnut, cypress, maple, beech, tamarack and hemlock.

Veneers

These are thin slices or sheets of wood with a uniform thickness. Although they may be cut as thin as 1/110th inch, the commercial product is usually about 1/20th in. thick, with 3/8th in being maximum. Veneers were known to the early civilizations of Rome and Egypt, but in America their use was retarded for many years because wood was so abundant and inexpensive.

To make veneers the logs or pieces of wood are peeled, boiled and then cut with a knife. The rotary process prepares most. This involves turning a log on a lathe against a stationary knife that produces a continuous sheet of veneer. The design in this type of veneer is not particularly striking because the sheet is cut parallel to the annual rings. In the slicing process the logs are quarter sawed and thus show a more attractive grain.

The logs are first quartered and then sliced with a stationary knife to yield separate sheets. This process is less wasteful and is used for the more expensive kinds of wood. In the sawing process the quartered logs are cut with a circular saw. Although the veneers thus produced are thicker, the most valuable woods are sawed because the fibres do not tare and the material can be more readily worked. Freshly cut veneers are usually wet and must be thoroughly dried before adhesives can be applied. Veneers were used primarily only to cover up inferior woods in the furniture and cabinet industries.

They were made primarily from walnut, mahogany and other woods that had a beautiful colour and grain. By the end of the 20th Century many species were being used and the veneers were utilized in the manufacture of baskets, boxes, cooperage, door panels, trunks, mirrors, musical instruments, etc. Veneers make possible high strength and minimum weight. Any wood is suitable that comes in large sizes, has a symmetrical grain and design and few defects and is inexpensive. A variety of both domestic and foreign woods are used, but more than one-half of the total output in North America is made from either Douglas fir or red gum.

There are three kinds of hardwood veneers manufactured. Face veneers are sliced or sawed from selected logs and are used for only the finest work. Primary among native species are black walnut, quartered red and white oak, red gum and sugar maple. Commercial veneers are rotary-cut and are used for plywood, concealed parts of furniture, etc. Birch, Maple, Beech, basswood, tulip, tupelo, cottonwood, sycamore and oak have all be used. Container veneers are the least expensive and are made from any inexpensive wood into barrels, crates, boxes, etc. Softwood veneers, either rotary-cut or sliced, are made primarily on the Pacific Cost. Over 80 percent are made from Douglas fir, but Sitka spruce, western yellow pine and Port Orford cedar have been used. Softwood veneers are for structural plywood or interior paneling.

Plywood

This involves gluing together 3-9 thin veneers. The grain of each successive layer is at an angle to the next, so the strength is redistributed and the dimensional instability of one layer is compensated for and reduced by the others. Thus, the finished produce is very strong and stable and much less likely to warp or twist than ordinary wood. Screws and nails may be driven close to the edge with no danger of splitting the plywood.

In the manufacture of a 5-ply plywood panel, the face, back, cross band and core sheets are prepared. Applying an adhesive and pressing the glued stock into a panel and finally drying and finishing the product follow this. Softer woods are usually used because they can be glued more easily. One simple kind of plywood has a 3/8^{ths} inch core of poplar with 1/10th inch birch veneers on each side. Various plant and animal adhesives are used as well as large quantities of synthetic resins. Modifications of the ordinary process result in molded or curved plywood and in wood alloys. Plywood is recognized as an engineering material with its own peculiar properties. It has extensive uses in the home for doors, flooring, walls, partitions, cabinets, shelves, furniture and interior trim. Large quantities have been used also to make concrete forms, prefabricated houses, airplanes, boats, railroad cars and the bodies of trailers and station wagons.

Wood Alloys

In making wood alloys or densified wood, such as uralloy, compreg and impreg, from plywood the natural wood structure is impregnated with synthetic resins and bonded under high pressure. The resins establish strong physical and chemical bonds with the wood fibres and create a material with new properties, which is also very strong, stable, hard and resistant to decay.

Reconstructed Wood

This is characterized by a reorganization of the fibres where they are taken out of their original unidirectional grain and rearranged in multidirectional patterns. This may be accomplished by chemical or physical means. Because the use of chemicals involves breaking down the wood into cellulose and lignin, the discussion of a chemically reconstructed wood, such as paper, is discussed under Sugars, Starches & Cellulose Products.

An example of a wood that has been reconstructed by physical means is Masonite. Wood chips are subjected to high pressure in a steam vessel and are then exploded with the abrupt release of the pressure. This tears the fibres apart and also reactivates the lignin, a natural plastic, which fixes and binds the fibres in their new orientation. At first Masonite was used as an insulating fibreboard because it was an excellent insulator against heat, sound and electricity. Then it was learned that the application of heat and pressure converted the boards into a homogeneous grainless synthetic board with extreme hardness and water resistance.

Shingles & Shakes

Wood Shingles

Shingles are thin pieces of wood used to protect the roofs and sides of buildings from weathering. Handmade shingles rank among the first of the wood products manufactured by humans. Single wood must be durable, light in weight, easy to split and able to hold nails without loosening, and it must not warp. Straight, even-grained woods are preferred. The durability of shingles is increased by treatment with a preservative. Red cedar has been the preferred wood in North America. Northern white and southern white cedars, redwood and cypress are also used. If available, eastern white pine is an excellent material. By the 21st Century many substitutes entered the market that were longer lasting and less fire prone. However, the lighter weight of shingles make them less expensive to install on existing roofs and aeration is superior to synthetic products, especially in warmer climates.

Wood Shakes

Shakes are split shingles and much thicker. They were important in colonial times, but now are used primarily for special architectural effects and in more remote regions. In North America red cedar, sugar pine and redwood are widely used.

Excelsior

This consists of thin curled strands or shreds of wood and is made by placing wood on frames and pressing it against rapidly moving knives or steel teeth. The material was first known as wood fibre. Excelsior is light and elastic and makes an excellent material for packing and shipping glassware and other breakable articles. It is resilient and free from dust and dirt, and may be used for stuffing upholstery and mattresses. A very fine grade of excelsior, wood wool, is used in filters and in the manufacture of matting and rugs. The use of excelsior has steadily decreased as synthetic plastics became available as packaging materials.

Sawdust & Shavings

Sawdust is used primarily for fuel, usually in the form of briquettes or fireplace logs. It is also valuable for many industrial purposes. It serves as a bedding for kennels and stables, as a floor covering to absorb moisture, for cleaning, drying and polishing metal, as a packing medium, as a soil conditioner and an insulating material. It has played a part in the making of leather and the conditioning of fur and is an ingredient of composition flooring, artificial wood, abrasives, wallboard, floor-sweeping compounds, etc. Shavings find their greatest use as fuel and as packing material. White pine shavings are especially desirable.

Wood Flour

This consists of finely ground sawdust, shavings and other forms of wood waste that has been used in the manufacture of linoleum, plastics, nitroglycerin, veneer bonds, composition flooring, insulating brick, etc. It is used as filler, an absorbent or a mild abrasive. Light coloured woods with low resin content are used. White pine has been the main species.

Charcoal & Wood Distillation

Wood Charcoal

Wood has been heated in order to convert it into carbon or charcoal since ancient times. It was probably the first chemical process used by humans. Charcoal is still a valuable fuel as it has twice the heating power of wood and burns without flame or smoke. It is widely used in many European, Asian and Latin American countries, especially where forests are abundant and is the chief domestic fuel in most tropical countries. Charcoal is also used in medicine, as a reducing element in the iron and steel industry, and in the manufacture of chemicals, explosives, gunpowder and some cosmetics. It is the best material for absorbing impurities and foul odors from both water and the atmosphere. It has been used extensively in gas masks.

The best yields of charcoal are obtained from the denser hardwoods, such as maple, beech, birch, oak, hickory and mesquite. Willow charcoal is especially popular for explosives. The conversion of wood into charcoal was formerly accomplished in open-air pits by a process of partial combustion. This method is wasteful as all the volatile material contained in the wood is lost. Beehive kilns and portable ovens have also been used. By the 21st Century charcoal burning has been largely replaced by wood distillation, and the valuable gases and other by-products are recovered as well as the charcoal.

Wood Distillation

This process is very old and was known to the ancient Egyptians. It is today important not only in rendering available the volatile wood elements, but as a factor in forest conservation. One of the main sources of wood for destructive distillation is the waste left by lumbering operations, sawmills, and planning mills. There are two distinct types of wood distillation.

Hardwood Distillation

This process uses the denser and heavier hardwoods and has been in use since the beginning of the 19th Century. The wood is heated in large oven retorts. The immediate products are charcoal, pyroligneous acid, which condenses from some of the gases given off, tar and oil, and noncondensable wood gases. The tar and oils are allowed to settle out and the pyroligneous acid is passed through a series of stills where more tar and oils are removed. Eventually slaked lime is added and in a final distillation wood alcohol (methanol) passes off and acetate of lime is left as a residue.

A modification of the process results in the recovery of acetic acid directly from the pyroligneous acid. The average yield per cord is 900-1080 lbs of charcoal, 180-200 lbs. Of acetate of lime or 103-125 lbs. Of acetic acid, 9.5-11 gallons of wood alcohol, 22-25 gallons of wood tar and oils, and 7,000-11,500 cubic feet of wood gas. Acetate of lime is used primarily in the manufacture of acetic acid, which has wide applications in the paint, textile, leather, film and plastics industries, and acetone, which is extensively used as a solvent. Wood alcohol finds it greatest use as a solvent, especially in the varnish and paint industry; it is the source of a variety of chemical products as well, such as aniline dyes and formaldehyde. It has a variety of uses as a fuel, illuminant, denaturant and ingredients of medical, chemical and other industrial preparations. The wood tar and oils are used as a fuel or as the source of many industrial oils. Wood gas also serves as a fuel and may be converted into a substitute for gasoline.

Softwood Distillation

This utilizes resinous woods, mainly southern yellow pine. The products are charcoal, wood turpentine, oils, and tar and wood gas. Softwood pyroligneous acid contains only small amounts of wood alcohol and acetic acid. The wood turpentine is used to some extent in the manufacture of varnishes, paints and synthetic camphor. The tar and oils also have industrial uses, one important product being creosote.

Misc. Uses For Wood

Wood is also used as a raw material for the paper and textile industries and as a source of tanning and dye materials, food, alcohol and other products, which is discussed in other sections.

Cork

This is a forest product of great antiquity. It is obtained commercially mainly from the cork oak, *Quercus suber*, a tree native to the Mediterranean region. This oak varies from 20-60 ft. tall and about 4 ft. in diametre, with a short trunk and densely spreading crown.

The evergreen leaves resemble those of holly, but are velvety and spongy. The acorns are used to feed pigs. The cork oak ranges from the Atlantic to Asia Minor and is especially numerous in Spain, Portugal, Algeria, Tunisia, southern France, Italy, Morocco and Corsica. The tree thrives on rocky siliceous soil on the lower slopes of mountains. Cork or corkwood consists of the outer bark of the tree, which can be harvested without injury to the tree. It is renewed annually. Harvesting consists of making vertical and horizontal cuts with hatchets or saws, and then prying off large pieces of the bark. The rich dark-red colour of the exposed areas is one of the typical sights in a cork forest that is being used for commercial purposes. The stripping is usually done in midsummer when weather conditions are favourable.

The bark of both the trunk and larger branches is usually used, although in some countries the cutting area is restricted to the first 6 feet of the trunk. Cork is first removed when the trees are about 20 years old. This first yield, which is known as virgin cork, is very rough and coarse and of little value. Subsequent strippings occur every nine years. The second yield is better, but the best quality of cork is not obtained until the third cutting and thereafter. The trees live for 100-500-years.and give an average yield of 40-500 lbs per tree. The best grade of cork consists of inch-thick layers obtained from young vigorous trees.

The stripped pieces of cork are dried for several days and weighed and then shipped to a processing area. They are then boiled in large copper vats. This removes the sap and tannic acid, increases the volume and elasticity, and flattens the pieces. It also loosens the outermost layer, which is scraped off. The rough edges are trimmed and the flat pieces are sorted and baled. Cork has many properties that make it valuable in industry. Despite its bulk, it is very light and exceedingly buoyant due to the fact that it is composed entirely of dead watertight cells.

It can be readily compressed and is very resilient. Even after 10 years of use cork stoppers can recover 7s5 percent of the original volume. It is durable, a low heat conductor, and is resistant to the passage of moisture and liquids. It also absorbs sounds and vibration and has certain frictional properties.

Uses of Cork

A great variety of products are manufactured from cork. Sometimes the natural cork is utilized; in other cases composition cork, made of coarse or finely ground pieces treated with adhesives and molded. Articles that have been made from natural cork include bottle stoppers, hats and helmets for use in tropical areas, tips for cigarettes, carburetor floats, handles for golf clubs, penholders, fishing rods, mooring buoys, floats, life preservers, life jackets, surf balls, baseball centres, decoys, mats, tiles, etc. Corkboard, made by heating natural cork, is used as an insulating material for houses, cold-storage plants, and refrigerators. It serves as a means of improving the acoustics of rooms and rendering them soundproof. It is very resilient and thus is a valuable material for machinery isolation.

Composition cork is used for the lining of crown caps, the metal tops for sealing bottles, gaskets, toes, counters, and innersoles for shoes, polishing wheels, friction rolls, and several types of floor covering. Linoleum is made from cork or wood flour, linseed oil, resins, such as rosin or kauri gum, pigments, and burlap. The oil is boiled and allowed to solidify by dripping on pieces of cloth. The solidified oil is ground up and melted with resins. This mixture is cooled and hardened, and after several days of curing it is mixed with the work, which has been ground to a fine dust, and with the dry colour pigments.

It is then pressed into burlap cloth with hydraulic presses. The linoleum is then seasoned in ovens and finished by giving it a protective surface of nitrocellulose lacquer. Linotiles are individual tiles made from ground cork and linseed oil, but much thicker and denser than linoleum.

6

Fibres and Fibre Plants

Plants yielding fibres have been second only to food plants in their usefulness to humans and their influence on the furthering of civilization. Primitive humans in their attempts to obtain the three most important necessities for life: food, shelter & clothing, focused on plants. Even though animal products were available, some form of clothing was needed that was lighter and cooler than skins and hides. It was easier to obtain from plants such items as bowstrings, nets, snares, etc. Also plant products were available from the leaves, stems and roots of many plants to construct shelter.

Very early on plant fibres have had a more extensive use than silk, wool and other animal fibres. Gradually as humans' needs multiplied, the use of vegetable fibres increased greatly until presently they continue to be of great importance even after the onset of plastics. It is impossible to estimate the number of species of fibre plants, but over a thousand species of plants have yielded fibres in America alone, and over 800 occur in the Philippines.

However, plant fibres of commercial importance ore relatively few, the greater number being native species used locally by primitive peoples in all parts of the world. Their durability often exceeds those of synthetic manufacture, one example being sisal & Manila hemps.

The most prominent fibres of the present are of great antiquity. The cultivation of flax, for example, dates back to the Stone Age of Europe, as discovered in the remains of the Swiss Lake Dwellers. Linen was used in Ancient Egypt and cotton was the ancient national textile of India, being used by all the aboriginal peoples of the New World as well. Ramie or China grass has been grown in the Orient many thousands of years.

Economic Classification (Fibres)

Plastic materials are often used instead of natural products because they cost less and sometimes tend to be more durable. However, natural plant products continue to have some superior attributes and are used when materials are readily available. There are six principal groups of fibres distinguished according to the way in which they are used.

Textile Fibres are the most important in that they are used for fabrics, cordage and netting. To make fabrics and netting flexible fibres are twisted together into thread or yarn and then either spun, knitted, woven or in some other way utilized. Fabrics include cloth for wearing apparel, domestic use, awnings, sails, etc., and also coarser materials such as gunny and burlap. Fabric fibres are all of some commercial value. Netting fibres that are used for hammocks, lace and all forms of nets include many of the commercial fabric fibres and a number of native fibres as well. Both commercial and native fibres are used for cordage. For this the individual fibres are twisted together instead of being woven. Binder Twine, fish lines, hawsers, rope and cables are among the many types.

Brush Fibres are stiff tough fibres including small stems and twigs that are utilized for making brooms and brushes.

Rough Weaving & Plaitling Fibres. Plaits are fibrous, flat and pliable strands that are interlaced to make straw hats, baskets, sandals, chair seats, etc. The most elastic strands are woven together for mattings and the thatched roofs of houses. The supple twigs or woody fibres are for making chairs, baskets and other wickerwork.

Filling Fibres are used for stuffing mattresses, cushions and in upholstery; for caulking seams in boats and in casks and barrels; as stiffening in plaster and as packing material.

Natural Fabrics are usually obtained from tree basts that are extracted from bark in layers or sheets and pounded into rough substitutes for lace or cloth. Fibres for Paper Manufacture includes textile fibres and wood fibres that are used in either the raw or manufactured state.

A plant cannot be restricted absolutely to any single group because the same fibre may be used for different purposes. Also, a plant may yield more than one kind of fibre. Thus the following discussion includes species that are considered in the group in which they are of the greatest importance.

Structure & Occurrence (Fibres)

All fibres are similar in that they are sclerenchyma cells that serve as part of the plant skeleton. They are predominantly long cells with thick walls and small cavities and usually pointed ends. The walls often contain lignin as well as cellulose. Fibres occur singly or in small groups, but they are more apt to form sheets of tissue with the individual cells overlapping and interlocking.

Fibres may occur in almost any part of a plant: stems, leaves, fruits, seeds, etc. The four main types grouped according to their origin include bast fibres, wood fibres, sclerenchyma cells associated with the vascular bundle strands in leaves, and surface fibres that are hair like outgrowths on the seeds of the plants. The term "bast fibre" is subject to criticism, as it gives no indication as to the particular tissue or region in which the fibres occur. It might be preferable to designate those fibres that occur in the outer parts of the stem as cortical fibres, pericyclic fibres or phloem fibres. But "bast" is a term that has been in use for a long time and is so established in commerce that it will be used in this discussion.

Fibres of economic importance occur in many different plant families, especially those from the tropics. Some of the more important families are the Palmaceae, Gramineae, Liliaceae, Musaceae, Amaryllidaceae, Malvaceae, Urticaceae, Linaceae, Moraceae, Tiliaceae, Bromeliaceae, Bombacaceae, and Luguminosae.

Textile Fibres

These fibres must be long and possess a high tensile strength and cohesiveness with pliability. They must have a fine, uniform, lustrous staple and must be durable and abundantly available. Only a small number of the different kinds of fibres possess these traits and are thus of commercial importance. The principal textile fibres are grouped into three classes: surface fibres, soft fibres and hard fibres, with the last two often referred to as long fibres.

Surface or short fibres include the so-called cottons. The soft fibres are the bast fibres that are found mainly in the pericycle or secondary phloem of dicotyledon stems. Bast fibres are capable of subdivision into very fine flexible strands and are used for the best grades of cordage and fabrics. Included are hemp, jute, flax and ramie.

Hard or mixed fibres are structural elements found mainly in the leaves of many tropical monocots, although they may be found in fruits and stems. They are used for the more coarse textiles. Sisal, abacá, henequén, agaves, coconut and pineapple are examples of plants with hard fibres.

Surface Fibres

Cotton is one of the greatest of all industrial crops. It is the principal fibre plant as well as one of the oldest and most economical. It was known since ancient times and well before written records. There are references to cotton by the ancient Greeks and Romans. Cotton was found in India before 1,800 B.C. The Hindus were believed to be one of the first people to weave cloth in the Eastern Hemisphere, although reference to Nordic traders of woven goods in North America during the Bronze Age has been made by Fell 1982. Cotton was introduced to Europe by the Arabs who called the plant "qutn." The plant had several origins because Columbus found it in cultivation in the West Indies, and it was known to the Amerindians of Neotropical America in Pre-Columbian times. Cotton became a commercial crop in the united States after 1787..

Cotton Characteristics

Several species of the genus *Gossypium* provide what we call cotton. The fine fibrous hairs that occur on the seeds constitute the raw material. These hairs are flattened, twisted and tubular. They compose the lint, floss or staple. Their length and other qualities vary with the different varieties. The plant is a perennial shrub or small tree naturally, but under cultivation it is treated as an annual. It branches freely and grows to a height of 4-8 ft. Cotton thrives in sandy soil in humid regions that are near water. This environment is typified in the southern United States and in the river valleys of India and Egypt. Cotton matures in 5-6 months and is ready to harvest soon after.

Kinds of Cotton

Hundreds of varieties have been developed from wild ancestors or produced by breeding during the long period of cultivation. Varieties differ in fibre character as well as other morphological features. Cotton is a difficult group to classify and the exact number of species is subject to argument. Cultivated cottons of commercial importance are usually referred to one or another of four species: *Gossypium barbadense* and *G. hirsutum* in the Western Hemisphere and *G. arboreum* and *G. herbaceum* in the Eastern Hemisphere.

Gossypium barbadense probably originated in tropical South America. The flowers are bright yellow with purple spots. The fruit, or boll, has three valves, and the seeds are fuzzy only at the ends. Two distinct types exist:

Sea-Island Cotton. This type has never been found in the wild as it was already being cultivated at the time of Columbus. It has fine, strong and light cream-coloured fibres that are regular in the number and uniformity of the twists and they have a silky appearance. These characteristics are valuable and sea-island cotton was formerly in great demand for the finest textiles, yarns, and lace and spool cotton. Sea-island cotton was brought to the United States from the West Indies in 1785.

The finest types were developed on the islands off the South Carolina coast and adjacent mainland. Here strong and firm stables of two inches or more in length were produced. Another form of sea-island cotton was grown along the coast in Georgia to Florida and in the West Indies and South America.

This has a staple of 1.5-1.75 in. in length. The yield of sea-island cotton was lower than other kinds of cotton, but this was compensated for by the greater value of the fibre. The boll weevil almost completely eradicated production of sea-island cotton before control measures were discovered.

Egyptian Cotton This cotton is grown in the Nile basin of Egypt where it was introduced from Central America. The plant is similar in appearance to sea-island cotton and is believed to be a hybrid.

However, the staple is brown in colour and shorter. Its length, strength, and firmness make this cotton suitable for thread, undergarments, hosiery, and fine dress goods. Egyptian cotton was brought to the United States in 1902 as an experimental crop and 10 years later it was recommended to farmers in the semiarid regions that were under irrigation. It was then grown in the western states of California, New Mexico and Arizona. Repeated selection and breeding resulted in the development of new strains of which Pima Cotton is of highest quality.

Gossypium hirsutum is a native American species that was grown by Pre-Columbian civilizations. It is usually called Upland Cotton, and is the easiest and most economical kind of cotton to grow. It constitutes the greater part of the cultivated cotton of the world. The flowers are white or light yellow and unspotted.

The bolls are four- or five-valved, and the seeds are covered with fuzz. Upland cotton thrives under a variety of conditions but does best in a sandy soil with abundant moisture during the growing and fruiting season and dryness during the time of boll opening and harvest as well as a temperature of 60-90 deg. Fahrenheit. The northern limit of economic growth is 37 deg. N. Lat. The Cotton Belt of the southern United States grows mostly upland cotton. The fibres are white with a wide range in staple length (5/8ths to 1.3/8ths in.).

There are over 1210 named varieties, many of which were developed through breeding experiments. The species probably originated in Guatemala or southern Mexico and spread northward to its present limits in North America. A well-marked variety, often recognized as a distinct species, occurs in the West Indies and along the dry coastal areas of South America as far as Ecuador and Brazil. Another variety occurs in Central America, northward along the Gulf of Mexico to Florida and the Bahamas and in the Greater Antilles.

Gossypium arboreum is the perennial tree cotton of Africa, India and Arabia. It was most likely the first to be used commercially, but production is now confined to India. The staples are coarse and very short (3/8ths to 34 in. long), but they are strong.

Gossypium herbaceum is the principal cotton of Asia. It was grown in Indian in ancient times and continues to be used locally there and in Iran, China and Japan. Its chief use is for fabrics, carpets and blankets and is often blended with wool. There are additionally several wild species of *Gossypium* in some tropical and subtropical areas.

Cotton Industry

Cotton used to be an expensive material because it was difficult to remove the fibres from the seed. The cotton gin developed by Eli Whitney in 1793 changed this situation and a revolution of the industry was started. Cotton then assumed a very prominent position in world commerce. The economics of cotton has had a profound effect on both the producing and purchasing nations. It is well accepted that slavery was perpetuated in America because of this crop.

There are several steps necessary in the preparation of raw cotton fibre in order to prepare it for the textile industry. These operations involve ginning in either a saw-tooth or a roller gin, baling, transporting to the mills, picking to remove any foreign matter and delivers the cotton in a uniform layer, lapping where three layers are combined into one, carding, combing and drawing where the short fibres are extracted and the others straightened and evenly distributed, and finally twisting the fibres into thread.

Cotton Uses

Cotton is used either by itself or in combination with other fibres in the manufacture of all types of textiles. Unspun cotton is extensively used for stuffing purposes. Treating the fibres with caustic soda, which imparts a high luster and silky appearance, makes Mercerized Cotton. Absorbent Cotton consists of fibres that have been cleaned and from which the oily covering layer has been removed. It is almost pure cellulose and makes up one of the basic raw materials of various cellulose industries.

A noteworthy advance in the cotton industry was the utilization of what were formerly waste products. In the early stages of the industry the cotton seed along with its fuzzy covering of short hairs or linters was discarded. However, all parts of the plant are now conserved to yield products that are valuable. The stalks contain a fibre that can be used to make paper or fuel and the roots possess a crude drug. The seeds are used for oil extraction and for livestock feed.

The linters give wadding, stuffing for pads, cushions, pillows, mattress, etc; absorbent cotton; low grade yarn for twine, ropes and carpets; and cellulose. The hulls are also livestock feed; fertilizer; lining oil wells to prevent cave-ins of the sides; as a source of Xylose, a sugar that can be converted into alcohol or various explosives and industrial solvents. The kernels yield an important fatty oil, cottonseed oil; and oil cake and meal are used for fertilizer, livestock feed, and flour and as a dye.

Soft or Bast Fibres

Flax

Once the most valuable and useful of fibres, flax gradually became less important as synthetics and cotton assumed more prominent roles. Flax is more durable than cotton and can yield a very fine fabric. The plant has been under cultivation for so long that its point of origin is unknown. It was used by the Swiss Lake Dwellers and was known to the ancient Hebrews and is frequently noted in the Bible. The ancient Egyptians wore linen and used it for the burial cloths. They carved pictures of the flax plant on their tombs. Long before the Christian era the Greek imported flax, and it is believed that the plant was being cultivated prior to 3,000 B.C.

Flax is in the genus *Linum* that contains several wild species of no economic importance as well as *Linum usitatissimum*, the source of the commercial fibre. The plant is an annual herb with blue or white flowers and small leaves. It grows to a height of from 1-4 ft. The fibres are formed in the pericycle and are made up of very tough, stringy strands from 1-3 ft. long that are aggregates of many long pointed cells with very thick cellulose walls. Flax does best in soil that is rich in organic matter and moisture and in temperate regions, but it may be grown elsewhere.

Preparation of the fibres is a more expensive procedure than for cotton. The crop is harvested and a process known as rippling breaks the stems. The fibres may then be rotted out by submerging the stems in water or by exposing them to dew. During this process called retting and enzyme dissolves the calcium pectate of the middle lamella, which holds the cells together, and frees the fibres. After retting the straw is dried and cleaned and the fibres are completely separated from the other tissues of the stem by an operation known as scutching. Finally the shorter fibres that constitute the tow are separated from the longer fibres. The long fibres are the only ones suited for spinning.

The fibres of flax have great tensile strength, staple length, durability and fineness. They are used in the manufacture of linen cloth and thread, canvas, duck, strong twine, carpets, fish and seine lines, cigarette paper, writing paper and insulating materials. Fibres from the stalks of flax grown for seed are too harsh and brittle for spinning but may be used for other purposes.

The principal production area was Northern Europe, with Russia producing around 70 percent of the world crop. Some of the finest flax is grown in Belgium. The Pilgrims introduced flax into North America and these and other colonists were growing sufficient amount for domestic use until 1900. Flax is a good crop with which to reclaim native soil and for a long time its cultivation was confined to the frontier. Flax is grown for its seed in areas with low rainfall. The seed is used in medicine and as a source of linseed oil.

Hemp

The term "hemp" is applied loosely to include a number of very different plants and fibres. The true hemp is *Cannabis sativa*, a plant native to Central and Western Asia but has spread worldwide where it often occurs as a troublesome weed.

The plant is a stout, bushy, branching annual that varies from 5-15 ft in height. It is dioecious with hollow stems and palmate leaves. The best grade of fibre is obtained from male plants. Hemp requires a mild humid climate and a rich loamy soil with an abundance of humus. Calcareous soils are especially suitable.

The fibre is white bast that develops in the pericycle. It is valuable because of its length that varies from 3-15 ft, its strength and great durability. However, it lacks the flexibility and elasticity of flax because of its lignification. Yields are usually high with one acre producing 2-3 tons of stems, 25 percent of which is fibrous material. The plants are harvested and shocked and dried. The fibres are separated from the rest of the bark by retting, either in dew or in water. They are then broken, scutched and hackled. Hemp must be harvested when the male flowers are in full bloom or the fibres are too week or too brittle to be of value.

Hemp is an ancient crop that had been grown in China before 2,000 B.C. It was introduced into Europe around 1,500 B.C. It reached North America in early colonial days and became a viable industry in Kentucky and Wisconsin. By the 21st Century very little of the crop was being grown in North America.

Hemp has been used to make ropes, carpets, twine, and sailcloth, yacht cordage, binder twine, sacks, bags and webbing. The waste and woody fibres of the stem were sometimes used to make paper. The finer grades can be woven into a cloth that resembles coarse linen. The short fibres or tow and ravelings constitute Oakum. This is used for caulking the seams between the plants in shipbuilding, in cooperage and as packing for pumps, engines, etc.

In the tropics hemp is grown for its seed, and also for a drug that is gotten from the flowering tops and leaves. The seeds contain oil that is useful in the soap and paint industries as a substitute for linseed oil. The drug, known as Hashish, is a resinous substance that contains several powerful alkaloids. In America this type of hemp is known as Marijuana. Ganja is a specially cultivated and harvested grade of hemp used for smoking and in beverages and candies. It has high resin content.

Jute

Jute has been used almost extensively as cotton even though it is much less valuable than either cotton of flax. It is a bast fibre obtained from the secondary phloem of two species of *Corchorus* of Asia. The best quality is from *C. capsularis*, a species with round pods that is grown in lowland areas subject to flooding. The plant is a tall, slender, somewhat shrubby annual with yellow flowers that grows to a height of 8-10 ft. It requires a warm climate and a rich, loamy alluvial soil. Fibre from *C. olitorius*, and upland species with long pods, is somewhat inferior but the two are not separated in commerce.

Harvest occurs within 3-4 months after planting and while the flowers are still in full bloom. The stems are retted in pools or tanks for several days to rot out the softer gummy tissues, and whipping the stems on the surface of the water then loosens the jute, or Gunny, strands.

The pale-yellow fibres are very long, from 6-10 ft. in length, and they are very stiff being highly lignified. They have a silky luster. They are produced in abundance, but are not especially strong and they tend to deteriorate when exposed to moisture. Despite these disadvantages they are economical and easily spun. Plastics have replaced many of the products formerly made from jute, however. Jute has been used mainly for rough weaving into burlap bags, gunnysacks and covers for cotton bales. The fibre is also used for twine, carpets, curtains and coarse cloth. Short fibres and pieces from the lower ends of the stalks make up jute butts that have been used in paper manufacture. India has the largest acreage of jute.

Ramie

Baehmeria nivea is a perennial, herbaceous or shrubby plant without branches when cultivated. It has slender stalks that reach a height of 3-6 ft. and they bear heart-shaped leaves that are green above and white beneath. The plant is from Asia and was grown in China in ancient times. It requires a fertile, well-drained soil. Several crops per year compensate for a rather low yield, especially in North America.

Fine fibres are obtained from the bast, which are very long, strong and durable. They also have a high degree of luster and would be desirable for textile purposes were it not for difficulties in the extraction and cleaning process. The stems are first immersed in water. The bark is then peeled off and the outer portions and green tissue are scraped off or are removed by oiling or mechanical means.

The fibres that remain are heavily coated with gum and require further treatment before they can be used. They make up the China Grass, or Filasse, that is used in the manufacture of grass cloth and other dress goods in Asia. Ramie has been used in Europe for under garments, upholstery, thread and paper. Although it is one of the strongest fibres known, being three times as strong as hemp, ramie has not been generally used because the treatment necessary to remove the fibres is very costly.

The development of a simpler process has not increased the use of ramie. Another variety, *Boehmeria nivea var. tenacissima*, is sometimes called Rhea. This plant is native to Malaya and resembles ramie except that the leaves are green on both sides. Rhea fibre is included under ramie in industry.

Sunn Hemp

Crotalaria juncea is an important fibre plant in Asia. It has been cultivated since ancient times and there are no known wild ancestors. It is the earliest fibre to be mentioned in Sanskrit writings (Hill 1952). It is a shrubby annual legume from 6-12 ft tall with bright yellow flowers. It is grown primarily in southern India.

Misc Soft Fibres

Almost all the Members of the Malvaceae Yield Bast Fibres That can be Used in Textiles. Some of the Most Important are as Follows: China Jute or Indian Mallow (*Abutilon theophrasti*) is an annual plant that yields a strong, coarse, grayish-white lustrous fibre with characteristics similar to jute. It has been extensively grown in China and was introduced into North America where it can thrive. The fibres have great tensile strength, take dyes readily and are used in China for making rugs and paper.

Kenaf (*Hibiscus cannabinus*) is a tall herb that yields a fibre with has over 129 names, among them Deccan, Ambari or Gambo Hemp, Java Jute and Mesta Fibre. It is a substitute for hemp and jute in the manufacture of coarse canvas, gunnysacks, cordage, matting and fishing nets. The plant is adapted to a wide range of climates and soils. Harvest is right after the flowers come into bloom. The fibres are 5-10 ft. long and are usually extracted by retting. Kenaf seed yields up to 20 percent of edible oil on being refined.

Roselle or Rama (*Hibiscus sabdariffa*) is cultivated in India, Southeastern Asia and some Pacific islands as a substitute for jute and for its edible fruit. The light-brown fibres are silky, soft and lustrous. Roselle is adapted to well-drained fertile soil where there is a 20-in. rainfall. It grows fast and can be harvested 90 days after planting. Retting is accomplished in 10-12 days, and the fibres are easily slipped from the bark. The red fleshy calices and involucels surrounding the young fruits are acid and provide a sour relish. The juice is used for flavouring and in making jellies, jams and wine.

Aramina or Cadillo (*Urena lobata*) occurs as a weed in most tropical countries. It provides a yellowish-white fibre that is more durable than jute and is used as a substitute in some industries. It has been grown commercially in Cuba, Madagascar, Nigeria, the Congo and Brazil where it is made into coffee sacks.

Other malvaceous species that yield fibres but which are of minor importance include Okra, *Hibiscus esculentus*, Majagua, *Hibiscus tiliaceus*, and several species of the genus *Sida*. *Sida acuta* is an extremely easy plant to harvest and prepare and the fibres are twice as strong as jute.

The Amerindians used the bast fibres of different plant for their bowstrings, nets, etc. Colourado River Hemp, *Sesbania exaltata*, was widely used by the western groups, while Indian Hemp, *Apocynum cannabinum*, and Milkweed, *Asclepias syriaca*, yielded important fibres for eastern groups of Amerindians.

Hard Structural Fibres

Abacá or Manila Hemp

This is a premier cordage material that is obtained from several species of wild plantain or banana. *Musa textilis* is the principal source. It resembles the true banana but has narrow more tufted leaves and inedible fruits. The plant forms a clump of 12-30 sheathing leafstalks 10-20 ft. high with a crown of spreading leaf blades 3-6 ft long. The fibre is secured from the outer portion of the leafstalks. Mature stalks are cut at the roots and split open lengthwise. The pulp and the fibre strands are removed, and the strands are washed and dried.

Individual fibres are 6-12 ft. long, lustrous and variable in colour from white to light ocher. They are light, elastic, stiff and very strong, durable and resistant to both fresh and salt water. Therefore, the main use of abacá has been in the manufacture of high-grade cordage, especially marine cables. Plastic cables are frequently substituted but they are not as easy to manage on ships as true abacá. Other products made from this kind of hemp are binder twine, bagging, strong tissue paper, papier-mâché, wrapping paper and Manila paper for sacks. In Japan Manila hemp was used in making movable partitions in houses. The individual fibres cannot be spun, but strands of fibres are used to make the lustrous cloth known as Sinamay.

Musa textilis has been of commercial importance only in the Philippines although it grows in other Asian countries as well. The plant was known and used by the inhabitants of the region centuries before the arrival of the first European explorers early in the 16th Century. The first shipment was made to North America in 1818. From then until 1918 it was the main export of the Philippines, amounting to over 300-thousand pounds annually (Hill 1952). More recently sugar and sometimes copra have exceeded it in production.

Manila help requires a warm climate, fertile soil, shade, good drainage, abundant moisture and an elevation lower than 3,000 ft. Suckers and rootstalks propagate it. The crop is grown in small fields or on large plantations and matures in 18-36 months.

After several failures, abacá was successfully introduced into the Western Hemisphere in 1925 in Panama, but did not develop into a commercial item. However, with World War II the serious shortages of Manila hemp became a serious threat to the war effort and the United States government financed a project in several Central American countries. Machines for cleaning the leaves were devised and installed, and soon some 26,000 acres were being cultivated that produced 3-million pounds of fibre. At the end of the war the industry was soundly established in Costa Rica.

Agave Fibres

By the middle of the 20th Century agave fibres were next to cotton in importance in America. By 1952 their value sometimes amounted to over 36 million dollars per year. But due to labour costs and the availability of synthetic alternatives their production declined thereafter. These plants are stemless perennials with basal rosettes of erect fleshy leaves. The leaves contain fibres that are removed either by hand or machine. There are numerous species of local occurrence. They are very drought tolerant and flourish in dry sterile soils. Several kinds of commercial importance are discussed as follows:

*Henequén or Mexican Sisal (*Agave Fourcroydes*)*

Amerindian groups have used this native Mexican species since ancient times. By the mid 20th Century Cuba and the Yucatan Peninsula produced most of the crop. The leaves bear spines that make them difficult to handle. The light straw coloured fibre is scraped out from the leaf tissue. It is hard, elastic and wiry, measuring 2-5 ft. in length. It was used mainly for binder twine, lariats and durable mats. It is not suited for marine or hoisting cables, as it is heavy and weak. Tablemats constructed from this fibre have a beautiful luster and are resistant to stains. *Agave letonae* from El Salvador is a related species. Production declined dramatically by the 21st Century.

*Sisal (*Agave Sisalana*)*

This is very similar in appearance to henequén but the leaves bear few spines. Native to Mexico and Central America it was cultivated in Hawaii, the East and West Indies and in several parts of Africa. The plant is very drought resistant and will grow where other species fail.

Little cultivation is required. The coarse, stiff, light yellow to white fibres are removed, cleaned, dried and packed in bales for shipment. Synthetic fibres also largely replaced sisal by the 21st Century.

Istle

In ancient times there were several fibres used in Mexico under the names of Istle, Ixtle or Tampico Fibre. Three species of most importance are Jaumaveistle, *Agave funkiana*, Tula Istle, *A. lecheguilla*, and Palma Istle, *Samuela* carnerosana. Several species of *Yucca* were also grouped under the category of Istle. The fibres are obtained from immature leaves of wild plants. Although these fibres are shorter than those of sisal and henequén, they are very strong and durable. They were formerly used for brushes and as a cheap substitute for sisal and abacá to make bagging, twine and rope.

Maguey

Manila Maguey or Cantala, *Agave cantala*, is a species from Mexico that was introduced into India and Southeastern Asia. It was grown commercially in the Philippines, Java and elsewhere as a substitute for sisal. Mexican Maguey is obtained from different species of *Agave* and the fibres are valued only locally by the inhabitants of the region in which it grows. It was nevertheless a highly valued plant by Amerindians who used it to make the fermented beverages of Pulque and Mescal.

Mauritius Hemp

Leaves of the green aloe, *Furcraea gigantea*, are the source of Mauritius hemp. The plant is native to tropical America but is grown worldwide, where local inhabitants use its fibre. It has been grown commercially in Mauritius, Madagascar, St. Helena and South Africa, India, Venezuela and Brazil, where it is know as Piteira. The plant resembles an agave but has larger, less rigid leaves and a very long peduncle or flower stalk that can reach of height of 20-40 ft. The fibres are very long, 4-7 ft., and they are white, soft, very flexible and elastic. They are not as strong as sisal and are used either alone or in a mixture for making bags, hammocks, coarse twine and other cordage.

Several other species of *Furcraea* yield fibres of local importance in tropical America. Included are Fique, *Furcraea macrophylla*, of Colombia; Cabuya, *F. cabuya*, of Central America; and Pitre, *F. hexapetala*, of the West Indies and sometimes called Cuban Hemp.

New Zealand Hemp

Also called New Zealand Flax, *Phormium tenax* is from the leaves of an iris like plant. It is native to wet areas of New Zealand but has been transported throughout the tropics and temperate regions of the world. In North America it serves as an ornamental. The fibres are very long, 3-7 ft. in length, and have a high luster. They are softer and more flexible than abacá and are used mainly for towlines, twine and other forms of cordage and mattings, and sometimes for cloth.

Bowstring Hemp

Many species of the genus *Sansevieria* occur as wild plants in parts of tropical Asia and Africa. These bowstring hemps are herbaceous perennials with basal rosettes of sword like leaves that arise from a creeping rootstalk. The leaves contain a strong white elastic fibre that has been used since ancient times for mats, hammocks, bowstrings and other types of cordage. Wild plants are generally utilized but some species have been cultivated. The fibres are removed by hand or mechanically. Important species include *Sansevieria thyrsiflora* of tropical Africa, *S. roxburghiana* of India and *S. zeylanica* of Sri Lanka. Several species were introduced into North America among which is the Florida Bowstring Hemp, *S. longifolia.*

Coir

This is a term applied to the short, coarse and rough fibres that make up a large part of the husk of coconut fruits, *Cocos nucifera.* It is the only prominent fibre that is obtained from fruits. Unripe coconuts are soaked in salt water for several months to loosen the fibres. They are then beaten to separate the fibres that are then washed and dried. The product has varied uses. In tropical Asia and Pacific Islands it is the source of Sennit Braid that is used for cables, small cordage and hawsers. Coconut fibres are superior to all others for this purpose because they re very light and elastic and resistant to water.

Coir has also been used for brush bristles, doormats, sacks, floor coverings, some textiles, upholstery, and stuffing for the bearings of railroad cars and as a substitute for oakum. Sri Lanka has been the centre for commercial production. In Puerto Rico coir was used in horticulture as a substitute for peat.

Pineapple

Pineapple, *Ananas comosus,* is the source of fibres of great strength and fine qualities. They are shiny white, very durable and flexible and are not harmed by water. When grown for the fibre pineapples are planted closer together and develop longer leaves. The best fibres are gathered from leaves that have not attained their maximum length. Two-year old leaves are usually harvested and the fibres scraped out by hand, which is an expensive process. After drying and combing, the fibres are tied end to end and can be woven. In the Philippines Piña Cloth is one of the most delicate and expensive of fabrics made from these fibres.

Pita Floja

Aechmea magdalenae is a plant that resembles pineapple. It is native to the dry alluvial soils from southern Mexico to Ecuador. The long leaves have a fibre of high quality known as Pita Floja or Pita. These fibres are the basis of one of the most ancient and most important native industries in Oaxaca and have also been used in Central America and Colombia. The fibres are 5-8 ft. long, white or light cream coloured, lustrous, finer and more flexible than other hard fibres and with a high tensile strength. They are very resistant to salt water so they are used to make fish lines and nets. They are also used for sewing leather.

Caroá

This fibre is a substitute for jute. It is from *Neoglaziovia variegata* a bromeliad of the dry, hot arid areas of northeastern Brazil. The leaves yield a soft, flexible, white elastic fibre three times as strong as jute. Caroá is used for rugs, sacks, textiles, cordage, twine and paper.

Brush Fibres

Brushes, brooms and whisks are made from various vegetable fibres. These fibres need to be strong, stiff and elastic with a high flexibility. Sometimes whole twigs, fine stems or roots are used, or the fibres are secured from leafstalks. Several important brush fibres are as follows:

Piassava

A few species of palms that grow in tropical America and Africa are the source of brush fibres called commercially Piassava, Piassaba or Bass Fibre. These trees have leaf stalks or leaf sheaths that yield the stiff, coarse, brown or black fibres in making brushes for sweeping large areas such as sidewalks and streets

West African Piassava is obtained from a wine palm, *Raphia vinifera* that grows in profusion in the tidal bayous and creeks of Liberia and other parts of West Africa. The leafstalks are retted and the bundles beaten. The long fibres are used to make mats and brushes. A wine is fermented from the palm tree sap.

Brazilian Piassava is from two species of palm found in profusion in the lowlands of the Amazon and Orinoco regions. *Attalea funifera* is the source of Bahia Piassava. The fibres are wiry, stiff and brown and almost like bristles. They are removed from the swollen bases of the leafstalks with an ax. They have been used primarily for street-cleaning machine brushes because the fibres are very durable and retain their resiliency even when wet. Para Piassava fibres are formed on the margins of the leaf petioles of *Leopoldinia piassaba*. They are used not only for brushes and brooms but also for hats, baskets and ropes.

Some other coarse fibres such as Palmyra and Kittul Fibre are classed as piassava in commerce. Palmyra Fibre is from the Palmyra palm, *Borassus flabellifer*, of the East Indies. This palm is one of the most useful as all parts of the plant are used for some purpose. The fibres are made into twine, paper, rope and machine brushes. Kittul Fibre is finer, softer and more pliable. It is obtained from the leaf sheaths of the toddy palm, *Caryota urens*, of Sri Lanka and the East Indies. The black bristles are made into strong ropes or into soft brushes. They also are substitutes for horsehair and oakum.

Cabbage palm, *Sabal palmetto*, of coastal southeastern North America yields a valuable fibre called Palmetto Fibre. The highest-grade fibre is obtained from young leaf stalks that are still in the bud. Coarser fibres come from mature leaves or the bases of old leaf stalks surrounding the bud. There was one an industry in Florida that processed this fibre for use as a substitute for palmyra in brushes. Palmetto fibres are reddish tan in colour and 8-20 in. long. The bud of the palm is edible and the roots contain tannin.

Broomcorn

The sorghum, *Sorghum vulgare var. technicum*, differs from other sorghums by having a panicle with long straight branches. This influorescence or seed head is the "brush" that is made into brooms. A dwarf variety exists that furnishes fibre for whiskbrooms, while the normal sized variety is used for carpet brooms. Harvest is before the end of flowering season by cutting the stems a few inches below the head. The heads are sorted, threshed and dried. Another species, *Spartina spartinae*, is a native grass of the southern Coastal Plain from Florida to Mexico that has been used in combination with the sorghum. Brooms often consist of as much as 50 percent *Spartina* surrounded by broomcorn.

Broomroot

Broomroot or Zacaton, *Muhlenbergia macroura*, is used to manufacture cheaper brushes. The plant is a grass found from Texas to Central America, especially in the mountainous regions of Mexico. It is a perennial with tufted wiry culms and coarse roots. The roots are the plant part utilized. They are harvested year-round, washed, cleaned and dried. They are then cut from the tops, graded according to quality, length and colour and baled for export.

Rough Weaving & Plaiting Fibres

There are relatively few materials that are manufactured for plaited or coarsely woven articles. The raw materials include the rushes, stems of reeds, willows, bamboo, grasses, rattan and leaves and roots. They are used entirely or split. They are woven or twisted together in a simple manner and made into sandals, mats, hats, matting, screens, chair seats, baskets, etc.

Hat Fibres

In many parts of the Eastern Hemisphere, rice, barley, wheat and rye are grown for the purpose of making braids or straw plaits for hats. The plants are grown close together so that they will have few leaves, and they are harvested before they mature. The stems are split lengthwise before plaiting. The Leghorn Hats and Tuscan Hats of Italy are some of the best of the straw hats.

Panama Hats are made from the leaves of Toquilla, *Carludovica palmata*, a stem less, palm like plant that grows wild in the forests from southern Mexico to Peru. It has been cultivated in Ecuador and parts of Colombia. The Panama hat industry is concentrated in Ecuador. Young leaves are collected while they are still folded in the bud and treated with hot water. The coarse veins are removed and the plaits are separated and split lengthwise into slender strips that are slowly dried and bleached. They gradually roll inward forming fine cylindrical strands known as Jipijapa. The hats are woven by hand from these strands. About six leaves are necessary to make one hat. The best quality Panama hats are uniform and have a fine texture, are strong, durable and elastic and resistant to water. The Puerto Rican Hats are made from the leaves of the hat palm, *Sabal causiarum*.

Mats and Matting

In the Eastern Hemisphere commercial mattings have been made from several rushes, grasses and sedges. Usually the stalks or leaves are used alone, but they may be combined with cotton of hemp. Some of the species utilized are Chinese Mat Grass, *Cyperus tegetiformis*, and Japanese Mat Rush, *Juncus effusus*.

The Screw Pines, *Pandanus tectorius* and *P. utilis* are important in Southeastern Asia and Oceania for making mats. The leaves of these species are also used for sugar bags, cordage, hats and thatching.

Baskets

Baskets have been and are continuously being made from an array of plant species worldwide. Roots, stems, leaves and even woody splints have been used. Commercial baskets are usually made from rushes, cereal straw, osiers or willows, and ash or white oak splints. Sweet grass baskets are made from *Hierochloe odorata,* a common species in lowlands along the coast and Great Lakes. Another important source of basket fibre is the raffia palm, *Raffia pedunculata,* native to Madagascar. Strips of the lower epidermis of the leaves are the raffia of commerce. The fibre is so soft and silk like that it can be woven. It is especially useful as a tie material for nurseries and gardens.

Wickerwork

This includes chair seats, chairs, infant carriages, hampers and other light articles of furniture. Willows, rattan and bamboo are the main plants used. Rattan is obtained from several species of climbing palms, Calamus spp., that grow in the humid forests of the East Indies and other parts of tropical Asia. The stems of these plants are long, strong, flexible and uniform. They are used either entirely or as splits in Asia for furniture, canes, baskets and other items. A considerable quantity of rattan is exported for making furniture.

Bamboos occur in most tropical areas, but they are especially abundant in the monsoon regions of Eastern Asia. They are the largest of the grasses with woody stems that sometimes reach one foot in diametre and a height of over 10 feet. There are many species in the families *Arundinaria, Bambusa, Dendrocalamus, Gigantochloa, Phyllostachys,* and other genera. The stems are used for all kinds of construction in areas where these plants grow. Exported bamboo is used in the manufacture of furniture, fishing rods and implements of various kinds. Bamboo splits are made into baskets and brushes.

In the Western Hemisphere bamboos have not been extensively utilized. *Guadua angustifolia* is a species with very strong culms and has been used in Ecuador to make furniture and in house construction.

Filling Fibres

Many plant fibres have been used to stuff pillows, cushions, furniture, mattresses, etc. They are also used to caulk the seams of vessels, in the making of staff for buildings, as stiffening for plaster, packing for bulkheads and machine bearings, and for the protection of delicate objects during shipment. Synthetic materials frequently take the place of these long used products, but in some ways they retain some superiority. Surface fibres are commonly used for stuffing because their staples are too short to be spun and thus are not valued in the textile industry. Bast fibres are too costly, and hard fibres are frequently too stiff and coarse. The silk cottons are the most important source for stuffing.

Kapok

This is the most popular silk cotton and most valuable of all the stuffing substances. Kapok is the floss produced in the pods of the kapok tree, *Ceiba pentandra.* Originally confined to the American tropics, it is now found worldwide. It is an irregular tree, 50-100 ft. tall, with a buttressed base and weird growth habit. It grows rapidly and begins to bare when only 15 ft. tall. A mature tree can produce more than 600 pods and from 6-10 lbs. of the cottony fibres. Pods are clipped from the branches and opened. The floss is removed and the seeds separated by centrifugal force.

The floss is 1/2-1 1/2 in. long and whitish, yellowish or brownish in colour. It is very fluffy, light and elastic and is thus an ideal stuffing material for mattresses and pillows. The fibres have a low specific gravity. They are five times more buoyant than cork and are impervious to water. Therefore, kapok is valuable as a filling for life preservers, cushions, portable pontoons, etc. Its low thermal conductivity and its high ability to absorb sound make kapok an excellent material for insulating small refrigerators and for soundproofing rooms. It has also been used for the linings of sleeping bags, gloves for handling dry ice and in the tropics as surgical dressings. Kapok seeds have 45 percent fatty oil that is extracted and used for soap and food.

Kapok Substitutes

There are a number of other plants with seed hairs or floss that can be used as a substitute for kapok. The Red Silk Cotton Tree or Simal, *Salmalia malabarica*, is a very large ornamental tree. It supplies reddish floss known as Indian Kapok that has been important as a stuffing in India for centuries. The White Silk Cotton Tree, *Cochlospermum religiosum*, yields a fibre of some importance. This handsome tree is native to India but is now widespread in the world tropics. It is also one source of Kadaya Gum. Madar, *gigantea*, and the related Akund, *Calotropis procera*, are shrubs native to Southern Asia and Africa that produce a silk of some importance. Although inferior to kapok, this substance is often used in mixtures with kapok. The Pochotes of Mexico, *Ceiba aesculifolia, C. acuminata*, etc., yield a silk cotton almost equal to kapok in buoyancy and resiliency. Palo Borracho, *Chorisia insignis*, and Samohu, *Chorisia speciosa*, of South America yield large amounts of a glossy, white silk cotton with properties similar to kapok.

All of the milkweeds have silky hairs on their seeds and several species are a source of stuffing materials. Milkweed floss is one of the lightest materials. It is very buoyant and a perfect insulator. It was used during World War II as a substitute for kapok. The pods contain oil and a wax that may have future applications. Some species yield textile fibres. In North America, *Asclepias syriaca* and *A. incarnata* produce abundant floss. In the Neotropics, *A. curassavica* has some value.

Misc. Filling Fibres

There are innumerable plants and fibres that have use as filling materials. Included are cereal straw, cornhusks, Spanish Moss and Crin Végétal.

Spanish Moss, *Tillandsia usneoides*, is a conspicuous tree epiphyte in Southeastern North America. This is an excellent substitute for horsehair after it is processed. The plant is pulled from the trees with rakes or hooks, or it is collected from the ground or water. It is then fermented in order to rot off the gray outer covering and ginned to remove impurities. The prepared fibre is brown or black, lustrous and very resilient. It has been used in upholstery and for automobile cushions.

Crin Végétal, *Chamaerop humilis* a dwarf fan palm of Northern Africa and the Mediterranean region in which the leaves have shredded and twisted fibres. These have been used as stuffing material.

Natural Fabrics

Some trees have basts with tough interlacing fibres that can be extracted from the bark in layers or sheets and can then be pounded into rough substitutes for cloth. Tapa Cloth is one of these as it once constituted the main clothing in Polynesia and parts of Eastern Asia. The material is obtained from the bark of the paper mulberry, *Broussonetia papyrifera*. Strips of bark are peeled from the trunk and the outer coating is scraped away. After soaking in water and cleaning these strips are placed on a hardwood log and pounded with a mallet. Overlapping the edges and beating them together unite the individual strips. The finished product varies according to thickness from muslin like material to one of leather. Tapa cloth is frequently dyed. Similar bark cloths have been manufactured from different sources since antiquity. In South America the Amerindians used the Tauary, *Couratari tauari*, and other species of the same genus. In Mozambique the wild fig, *Ficus nekbudu*, was used as a source of Mutshu Cloth. The Upas Tree, *Antiaris toxicaria*, of Sri Lanka furnishes a bark cloth. It is also the source of an important poison used with arrows.

Lace Bark is the produce of Lagetta lintearia, a small tree of Jamaica. The inner bark is removed in sheets and can be stretched into a lacelike material with pentagonal meshes. It is suitable as a textile and ornament. Cuba Bast is from *Hibiscus elatus*, a small bushy tree of the West Indies. The inner bark is removed in long ribbon-like strips that have been used in millinery and for tying cigars.

The vegetable sponges, *Luffa cylindrica* and *L. acutangula*, yield a unique fibre. These are climbing cucumbers of the tropics that bear edible fruits containing a lacy network of stiff curled fibres. This material is extracted by retting in water. After cleaning it is used for making hats, for washing and scouring machinery, in certain types of oil filters and as a substitute for bath sponges. A large amount of this material used to be exported by Japan.

Papermaking Fibres

The manufacture of paper requires the use of cellulose present in plant fibres. This subject is discussed under Forest Products

Artificial Fibres

The artificial fibres in use in the textile industry are mostly organic in nature, with synthetic glass fibres being the exception. The organic materials utilized are cellulose, plant and animal proteins, and synthetic resins, such as nylon that is made from soft coal, water and air. The cellulose fibres are discussed under Forest Products, while the protein fibres of only minor importance. Although they have many of the general properties of wool, their low strength when wet is a serious detriment. Soybeans, corn and peanuts are the main plant sources of protein fibres.

7

Rubber and Other Latex Products

Rubber is acquired from the milky fluid, or latex, of various erect or climbing woody plants of the tropics or subtropics. Most of the rubber plants belong to the Moraceae, Euphorbiaceae, or Apocynaceae. Although more than 50 species are available as sources, only a few have been important commercially, with *Hevea brasiliensis* being the most prominent. Originally only wild trees were the source of rubber, but later cultivated *Hevea* trees, the "plantation rubber" furnished about 98 percent of the supply (Hill 1952).

Rubber is the most recent of the major crops of the world. The industry by 2003 was about 150 years old, but cultivation has been carried on only about 100 years. The increase in the production of plantation rubber was from 800 long tons in 1900 to 305,000 tons in 1920 and 1,395,000 tons in 1940. This development of the rubber-growing industry has not been without drawbacks. Overproduction had seriously affected the industry financially by 1950, and many attempts have been made to regulate its supply. The British and Dutch, in particular, tried to restrict production and exert other methods of control in Malaya, Java, Sumatra and other plantation rubber centres within their spheres of influence. The successful development of synthetic substitutes for rubber, after many years of experimentation, further jeopardized the natural rubber industry. Nevertheless, while these substitutes are superior for some purposes, such as the conduction of oil, natural rubber is still preferred for vehicle tires, which utilize about 3/4ths of the rubber output.

Latex occurs in special cells or in a series of special vessels, which permeate the bark, leaves, and other soft parts of the tree. Usually only the latex from the lower part of the trunk is of importance commercially. Latex is a gummy white liquid full of minute globules. It consists of a mixture of water, hydrocarbons, resins, oils, proteins, acids, salts, sugar and caoutchouc, the substance used as the source of rubber. The significance of latex to the plant is obscure, but it is know to be of some value in the healing of wounds, and it may serve for protection, nutrition, and the transport of materials or as a fluid reservoir.

The properties of rubber were known by primitive cultures in Central America and Peru. Their word *cauchuc* has been changed to the present caoutchouc (*caucho* in Spanish). Columbus was the first to report the existence of rubber to Europeans, but it was not introduced into Europe until 1735. Priestley first applied the name "rubber" in 1770, due to the fact that caoutchouc could be used for removing pencil marks. Mackintosh in 1823 developed a process for waterproofing cloth, but it was not until 1830 when Goodyear made the discovery of the vulcanizing process, that rubber really became popular. From then on the rubber industry developed rapidly, and today rubber has a vast and constantly increasing number of uses and industrial applications. The properties that make it so valuable include its plasticity and elasticity, its resistance to abrasion and to electrical currents, and the fact that it is impermeable to liquids and gasses.

Hevea Rubber

The Hevea or Para rubber tree (*Hevea brasiliensis*) is the normal source of 95-98% of the rubber produced throughout the world. The tree is native to the hot damp forests of the southern affluents of the Amazon River in South America. It was estimated in 1952 that that there are over 300-million trees in the vast area traversed by the Amazon and the Orinoco. Within this region the optimum conditions for its development are found. This includes a uniform climate with a temperature range from 75-90 deg. Fahrenheit and a rainfall of 80-120 inches.

The trees may reach an age of at least 200 years and may attain a height of 60-140 ft. The leaves are 3-lobed, and the flowers are small and inconspicuous. The fruits have three seeds, which contain 23-32 % of fatty oil. This oil is sometimes extracted and used as a drying oil. The resulting oil cake is rich in proteins and is used as a stock feed. Although rubber trees are natives of swampy floodlands, they grow best on deep, fertile, well-drained upland soils at an elevation of 600-1500 ft. above sea level. The latex occurs in a series of vessels in the cortex. Several other species of *Hevea* are tapped, including *Hevea benthamiana*, which has high-quality latex, but their production is insignificant compared with *H. brasiliensis*.

Wild Rubber

In the beginning of the industry only wild rubber trees are available as a source of rubber. The maximum production was reached in 1910 with an output of 83,000 long tons. After that with the development of plantation rubber, wild rubber production steadily declined to as low as 8,500 tons in 1932. Since that time the output has fluctuated, depending on price and demand. During World War II wild rubber was again actively exploited.

The methods employed in obtaining and preparing the latex have changed little over the years. The collection of the latex is in the hands of local residents, called "seringuieros" who are usually in the employ of absentee owners (Hill 1952). Each seringuiero is responsible for a single tapping circuit, which contains from 35-180 trees, about 2 or 3 to the acre. When a new tree is located, it is tapped at a point about 3 ft. above the ground by cutting several short downward 30-degree panelike incisions in the bark with a special knife.

The cuts are deep enough to sever the latex vessels but do not extend far enough to injure the cambium. Cups are placed below the incisions to receive the latex, which flows for several hours. The seringuiero makes the round of his trees each day. Successive tappings consist of entirely new incisions made slightly below the previous ones.

The latex is collected from the cups and carried to the camp for coagulation. This is done by coating a pole with latex and suspending it over a fire made of palm seeds or special woods. These yield a dense smoke containing acetic acid, creosote, and tars which coagulates the latex, forming a layer of crude rubber. The process is repeated until balls weighing 125-200 lb. are obtained. In some areas paddles are dipped in the latex and held in the smoke. The balls are shipped to mills for processing.

Plantation Rubber in the Eastern Hemisphere

Henry Wickham transported 70,000 *Hevea* seeds from the Amazon to England in 1876. These were grown at Kew, and enough seedlings were obtained to start a rubber industry in the eastern tropics, at first in British Malaya and Ceylon and later in Java, Sumatra and other areas (Hill 1952). There had been earlier attempts at rubber cultivation, but plantation rubber did not become permanently established until the beginning of the 20th Century. In 1910, when Amazon rubber production was at its peak, the East produced only 11,000 tons. Four years later it had outstripped its rival, and by the outbreak of World War II, it was responsible for 98% of the world's output of 1,500,00 tons, with nine million acres under cultivation.

The rapid development of plantation rubber in the East may be attributed to a favourable climate, absence of diseases, low labour costs, and the extensive research programme launched by the British. Every step in the production of the crop was rigidly supervised, and the best methods of planting, cultivation, tapping and coagulation were discovered. A selection and breeding programme facilitated an increase in yield. The discovery of bud grafting allowed that buds of high-yielding plants be grafted on ordinary seedlings.

The resulting clones differed greatly, and only those with the most desirable traits were retained. Using such approved clones began in 1925. By 1945 10% of the plantations had trees that were the result of grafting, and they obtained a yield of 1500 lb per acre. The 90%, which still had a random mixture of trees, had a yield of only 450-500 lb/acre. Where controlled crosses of high-yield clones were made, yields as much as 2000 lb. per acre were possible. Not all plantation rubber was produced on the large estates. The acreage devoted to rubber grown by small growers was produced in small family gardens and was actually more than that grown on estates.

Plantation Rubber in the Western Hemisphere

Following the imposition of export restrictions on eastern rubber in 1924, the United States favoured the establishment of plantations in other parts of the world. The Firestone Corporation selected Liberia, but Ford Motor Co. and Goodyear turned to tropical America, the native area of *Hevea*. The 2,500,00-acre Fordlandia was started in Brazil in 1928, while Goodyear chose Panama and Costa Rica. All the American efforts failed after a promising beginning, due primarily to the ravages of leaf blight, *Dothidella ulei*, to which plantation rubber in the Western Hemisphere was very susceptible.

It was obvious that rubber could not succeed as a crop in the American tropics without adequate disease control. A research programme did produce satisfactory results, however. An effective spray, of copper and sulfur fungicides with various spreaders and stickers, was developed and could be used on seedlings. Yield and resistance experiments were conducted on thousands of wild trees, with the most desirable being selected for propagation. Fortunately there was a successful importation of approved eastern clones from a

Goodyear plantation in the Philippines. When these were budded on resistant stock from the jungles, desirable combinations of characters were obtained. Double budding also proved satisfactory. This involves budding a high-yielding eastern clone onto a native root as soon as the plant is large enough. This is grown in a nursery under spray control until it is 6-10 ft. tall. Then the top is budded with a disease-resistant strain.

The resulting rubber tree has a native root system, a high-yielding eastern clonal trunk, and a disease-resistant crown. Long-term breeding programmes were continued well into the mid 1900's, as well as extensive explorations in the jungle, with the goal of obtaining strains that combined disease resistance and high yield and which would reduce the expensive spraying and double-budding practices. The cooperation of the United States with governments in Latin America resulted in rubber being grown successfully in many areas. Emphasis was placed on small family plantings rather than on large plantations. The rubber trees were usually set out at intervals of 12-24 ft. in tows 20 ft. apart. Other crops were grown between the rows.

Thus, from 1/3rd to 1/2 of the acreage was resting; the balance was tapped on some alternating programme. At each successive tapping a thin shaving of bark was sliced off the original panel until it almost reached the ground. One-third, 1/2 and full spirals were used. Following the collection of latex, a small amount of ammonia or some other anticoagulant was often added to keep it liquid until it reached the processing area, where it was concentrated and made into sheet rubber. Cleaning the latex and pouring it into large pans accomplished this. A coagulant, such as formic or acetic acid, was added, and in several hours a mass of soft rubber resulted. Blocks of this were washed and run between rollers to form thin sheets.

These were smoked and baled for shipment. Other types of crude rubber were also prepared. "Worm rubber" consists of irregular wormlike pieces cut from sheets with a pair of shears. "Crepe rubber" was made by passing washed and bleached coagulated latex through a creping machine, which turned out long, thin, perforated strips of rubber. In the preparation of "sprayed rubber" latex was dropped on whirling disks and little particles of rubber were thrown off. Any moisture quickly evaporated. This kind of rubber was exceedingly pure and clean. Also, steadily increasing amount of liquid latex was evaporated. This required immediate processing in the field and was feasible only for large plantations.

Castilla Rubber

Panama or Castilla rubber is from *Castilla elastica*, a tall tree native to southern Mexico and Central America. There are many names for this tree in the indigenous languages of the region. It was the chief source of rubber from 1794-1850. Other species of *Castilla* may also be used, among which are *C. ulei* of the Amazon, the source of caucho rubber. *Castilla elastica* is found in deep loamy soil on high ground and may reach a height of 150 ft. It requires temperatures above 60 deg. F. and requires shade when young. Often the natives cut these trees down in order to obtain the large possible yield of latex, which could ultimately lead to this species extinction.

Trees are tapped when 8-10 years of age, with adult trees yielding a maximum of 50 lb. of latex. This is coagulated with plant juices, alum and by boiling or exposure to the air. The crude rubber is shipped in flat cakes. Castilla rubber was once extensively cultivated in Central America in plantations, but it could not compete successfully with Hevea rubber. The plantations were abandoned. The surviving trees were the source of some of the emergency supply of Castilla rubber during World War II (Hill 1952).

Ceara Rubber

Ceara or Manicoba rubber is obtained from *Manihot glaziovii*, a small tree native to drier regions in Brazil. It grows well in dry rocky ground so that it may be utilized in areas unsuitable for other types of rubber. The tree grows rapidly, reaching its maximum height of 30 ft. in a few years. Ceara rubber is also grown in Ceylon, India and other tropical countries. The trees are tapped when 4-5 years of age and yield a good grade of rubber. The latex is coagulated by exposure to air or smoke. The crude rubber is exported as blocks or flat cakes.

Assam Rubber

Assam or India rubber as it is often called, is obtained from *Ficus elastica*, a native tree of northern India and Malaya. The tree requires a hot climate and a high rainfall. It often begins life as an epiphyte. It grows to a great height, developing huge buttresses or prop roots. The roots as well as the stem are tapped. The latex is allowed to drip onto bamboo mats, where it coagulates. A large amount coagulates on the trunk as well. This crude rubber is scraped off, cleaned and dried. The native procedures for tapping the wild trees have been very wasteful, so the tree is cultivated in some areas. The yield is low and the trees do not mature until about 50 years of age. Assam rubber is of low grade and has been of little or no commercial importance.

Lagos Silk Rubber

Lagos silk rubber is from *Futumia elastica*, a large tree of tropical West Africa. Discovered in 1894, it was immediately exploited to the point of near extermination. In 1898 over 6 million pounds were exported. The tree has been cultivated and an attempt made to reestablish the industry. However, plantations of Hevea rubber are the chief source of rubber in West Africa.

Landolphia Rubber

Other former sources of African rubber were several woody climbers in the genus *Landolphia*. The most important were *L. kirkii* on the east coast and *L. heudelotis* and *L. owariensis* on the west coast. The huge vines, sometimes 6 in. in diametre. Were pulled down and cut into small pieces. The latex that exuded was coagulated with plant juices or the heat of the sun, sometimes even on the bodies of the native collectors (Hill 1952). The most wasteful methods were employed, for a large immediate yield was desired. There was very barbaric treatment of the natives involved in this harvest. Operations in the Belgian Congo during the reign of Leopold II remain some of the most awful pages in history. Although there is still a considerable amount of wild landolphia rubber, there is apparently no future commercial value for it. Cultivation of this plant is not practical.

Guayule Rubber

The guayule, *Parthenium argentatum*, a native of America, has been utilized sine 1910 as a minor source of rubber. This low semi-shrubby plant grows in the arid regions of Mexico and the southern United States. The early Amerindians who obtained the caoutchouc by chewing the plants knew it. Unlike the other sources of rubber, there is no latex present in the guayule plant.There are small granules of caoutchouc scattered all through the tissues. These are extracted mechanically or by means of solvents. The entire plant is pulled up and chopped up or ground with water.

The particles of caoutehouc float to the surface and are removed. They are then dried and pressed into slabs. If solvents remove the resinous materials the product is as good as the best Hevea rubber. Guayule rubber is especially good for mixing with synthetic rubber substitutes. Guayule has been adapted to plantation culture and it was extensively investigated as part of the Emergency Rubber Programme during World War II.

The yield has been increased; the life cycle shortened, and improved cultural, harvesting, and processing machinery developed. Production costs are still high so that it does not compete with Hevea rubber except in emergencies. The importance of guayule for the future lies in the fact that it constitutes a living stockpile of rubber on land which otherwise would be idle (Hill 1952).

Dandelion Rubber

The dandelion of Russia, *Taraxacum kok-saghys*, is a close relative of the common dandelion. In 1931 Russian botanists discovered it while searching for new economic plants. Rubber is present in considerable quantities in latex tubes in the long taproots. The yield has been increased by selection and breeding, and the plant was grown on some 2 million acres during the time of the Soviet Union. Seeds were brought to the United States in 1942 and planted experimentally in 42 states. Vigorous plants with greater root weight and rubber content were segregated. Hill (1952) believed that a yield of 400-500 lb. per acre could be achieved through continued research. The plant is well adapted to the northern United States and Canada, where it can be harvested the year it is planted. Under the most favourable conditions it might be profitable to grow this species as a minor source of rubber, and indeed was cultivated for this purpose in Argentina.

Minor Sources of Rubber

Other species of plants have been considered as a possible source of rubber. Among these are intisy (*Euphorbia intisy*). This leafless shrub of arid regions of Madagascar contains latex that has had considerable use locally.

The rubber coagulates on the surface of the plant in long elastic strands and is of high grade. Attempts have been made to cultivate intisy in the United States because of its adaptation to desert conditions.

Cryptostegia grandiflora and *C. madagascariensis* are ornamental woody climbers from Madagascar. They are now found in the tropics and subtropics of both hemispheres and were seriously considered during World War II as a new source of rubber. Both species have been grown for many years in India as a source of high-grade palay rubber. They are abundant in Mexico, occuring there in a wide variety of soils and climates. They seem the hardiest and fastest growing of all rubber plants, attaining harvest age within six months. They seem to offer considerable future promise, but attempts in Haiti to grow them on a large scale failed due to difficulties of extracting the rubber.

Other rubber-bearing plants of local importance include the mangabeira (*Hancornia speciosa*), a shrub or small tree of Bolivia, Brazil and Paraguay; a species of *Micrandra*, which yields the caura rubber of Venezuela; and various species of *Sapium* in northern South America.

Also a few native North American plants have rubber-contining latex. Among those, which have been given some attention, are the desert milkweed (*Asclepias subulata*); the Indian hemp (*Apocynum scannabinum*); several goldenrods, in particular *Solidago leavenworthii*, which was investigated by Thomas A. Edison; species of the rabbit brush (*Chrysothamnus*), the source of chrysil rubber; and species of *Cnidoscolus* from which chilte rubber is obtained.

Other Latex Products

Gutta-Percha

Gutta-percha is a nonelestic rubber obtained from the grayish-white latex of several members of the Sapolaceae. It has been known since 1842. The chief source is *Palaguium gutta*, a tree of Malayan origin, which has been grown in Borneo, Sumatra, the Philippines and other tropical countries. The latex is produced in sacs, which occur in the cortex, phloem, pith and leaves. It is obtained by making incisions from which the milky juice runs out very slowly, or by felling the trees. The latter is the usual method. The bark is removed in strips one inch in with and one foot apart, and the latex is collected in coconut shells or in palm of plantain leaves. The latex soon coagulates into grayish-yellow masses of a hard substance, which is odorless and heavier than water. This crude product contains several resins and other impurities and is purified by washing in hot water.

The whole mass is boiled and then kneaded into blocks, or it is chopped or sliced up and the pieces are washed, strained and kneaded, and then rolled into thin sheets (Hill 1952). The value of gutta-percha depends on the amount of a hydrocarbon, gutta that is present. Gutta-percha is hard at ordinary temperatures. It deteriorates very rapidly when exposed to air through oxidation and should be kept under water. It softens at 77 degrees Fahrenheit, can be kneaded at 122 deg. F., and melts at 248 deg. F. Because it is an exceedingly poor conductor of electricity it has been used for insulation. For many years no other material could replace gutta-percha and the similar balata in the construction of submarine cables, which require a substance that is resistant to salt water, pliable and with just the right amount of rigidity. Other uses included splints, supports, pipes, golf balls, speakingtubes, telephone receivers, waterproofing and adhesives. It has also been utilized for protecting wounds and in dentistry.

Balata

Balata is a nonelastic rubber that is obtained from the latex of *Manilkara bidentata*, formerly known as *Mimusops balata*, and other species of the genus. It is a native of Trinidad and South America. This magnificent tree grows to a height of over 100 ft. When mature its purplish wood is very hard and durable and is much used for ties and building purposes under the name of bully wood or bulletwood. The fruit is also edible.

The latex is obtained by tapping the trees three times each year. It flows freely and readily coagulates in the air. A tree 3 ft. in circumference will yield 50-100 lbs. of dry balata. After coagulation it is cleaned and molded into cakes. Balata contains about 50% gum. It serves the same purpose in industry as gutta-percha. It is particularly well adapted for machine beltings as it grips tightly and never stretches. It is also used as a substitute for chicle. Balata has been known since 1859, but unlike gutta-percha, it has never been cultivated. Several inferior types of balata are found in the Amazon region, chief of which is abiurana or coquilana obtained from *Ecclinusa balata*.

Jelutong

Jelutong is obtained mainly from *Dyera costulata* and related species from Malaya. These trees have a copious flow of latex, greater than all the other latex species combined. From 1910-1015 this was exploited as a source of rubber. However, the latex is combined with many gums, resins and other impurities, and it yielded a poor quality of rubber. Jelutong has been used mainly as a substitute for chicle.

Chicle

Sapodilla or naseberry (*Achras zapota*) is a tall evergreen tree native to the Yucatan Peninsula of Mexico. It is cultivated in tropical America and Florida for its edible fruit. The bark contains latex, 20-25% of which consists of a gum, or chicle, similar to that of gutta-percha. It is the basis of the chewing-gum industry but is also used in making surgical tape and dental supplies. The primitive method of obtaining chicle involves tapping the trunk and then scraping the thickened exudates from the bark. The crude or leaf chicle consists of pink or reddish-brown pieces mixed with 25-40% impurities.

In southeastern Mexico and British Honduras where the industry occurs most abundantly, the native collectors or chicleros are more careful. Zigzag gashes are cut in the trunk with a machete, up to a height of 30 ft. The latex runs to the base of the tree where it is collected in rubberized bags, leaves or even hollows in the earth. This accounts for the grains of sand that are oftentimes found in chewing gum. The flow of latex lasts for several hours and the yield may be as much as 55 litres. In order to conserve the supply, plantations were established. However, these were not practicable because the trees can be tapped only every 2-3 years.

The hardened chicle is boiled; a process that requires considerable skill, for the chicle must be poured off when the moisture content reaches 33%. It is then molded into blocks for shipment. Raw chicle contains resin, gutta, arabin, calcium, sugar and various soluble salts. Purification involves breaking the blocks into small pieces, washing in a strong alkali, neutralizing with sodium acid phosphate, washing again and finally drying and powdering.

The final product is an amorphous pale-pink powder, which is insoluble in water and forms a very sticky mass when heated. Ageing partially oxidizes the mixture that then turns brown and becomes very brittle. The final steps in the manufacture of chewing gum involves cleaning, filtering, sterilizing and compounding with various flavouring materials. Thirteen pounds of the final product yields about 5000 pieces of chewing gum. A piece of gum usually contains about 15% chicle, the remainder being chicle substitutes, sugar and flavouring substances.

Attempts to find substitutes have not been very successful, but inferior latex from other sapodillas, balata and jelutong are all utilized to some extent. Sorva or leche caspi, obtained from *Couma macrocarpa*, a large tree of the upper Amazon, was exploited during World War II. The tree, which has very abundant latex, is felled, ringed and the latex is collected in cups made from palm leaflets. The latex is then coagulated by boiling and is shipped in the form of large blocks. The white wood of this species is also used for furniture, and the fruits have a mucilaginous but edible pulp.

By 1952 the United States was the greatest chewing-gum market and used about the entire output of chicle. The supply was chiefly from Mexico, Guatemala and British Honduras.

Sugars, Starches and Cellulose Products

Sugars

Green plants manufacture sugars so that they all contain some quantity of sugar. However, much of the manufactured product is used directly in plant metabolize that very little usually accumulates. Storage sugars are found in roots, as with beets, carrots, parsnips; in stems as in sugar cane, sorghum, maize and the sugar maple; in flowers, such as in palm trees; in bulbs like the onion; and in many fruits.

There are several kinds of sugar, principal among which are sucrose or cane sugar, glucose or grape sugar and fructose or fruit sugar. They all seem to serve as a reserve food supply for the plant.

Humans require sugar in their diet. It constitutes a perfect food, as it is a form that can be readily assimilated in the body. Its main value is as an energy producer, and it is especially well adapted for use after heavy exercise. A large industry has developed in connection with the extraction of sugar from plant tissues, purification and refining. Additionally over 10 thousand different chemical derivatives have been made.

Sugar is an especially valuable product derived from the plant world. Only wheat, maize, rice and potatoes surpass it in importance. Yet there are relatively few sources for this industry. Only the sugar cane, sugar beet, sugar maple, maize, sorghum and several palms are commercial sources. Sucrose is the type of sugar stored in all of these plant species.

Sugar Cane: Most sugar is derived from sugar cane, *Saccharum officinarum.* It is a vigorous and rapid-growing perennial grass reaching a height in cultivation of 8-12 ft and a diametre of about 2 in. It grows in clumps with bamboo like stems arising from large rootstalks and with very ornamental feathery plumes of flowers. The stem is solid with a tough rind and numerous fibrous strands, and contains about 80 percent juice, the sugar content of which varies considerably from area to area and season to season.

Commercial sugar cane is a cultivar that is not known in the wild state. The plant was most likely first domesticated in Southeastern Asia or the East Indies from some wild ancestor from that region. By 327 B.C. it had become an important crop in India. It reached Egypt in 641 A.D. and Spain in 755 A.D. Since that time sugarcane has gradually been introduced into most humid tropical and semitropical regions.

The Portuguese and Spaniards were great disseminators of the plant into the New World. They carried it to Madeira in 1420 and to America by the beginning of the 16th Century. Within another 100 years it had spread all over the West Indies, Central and South America. Sugar cane first arrived in the United States in Louisiana in 1741. The name "sugar" is derived from Sanskrit "sarkara," meaning gravel, and refers to the crude sugar, which was the only kind known for centuries.

Sugar cane has been the principal export crop of the tropics and is unaffected by many of the conditions that influence the growing of other crops. It will grow well in any moist hot region where the average rainfall is 50 in. or more per year and where there is abundant sunshine and where temperatures do not drop below 70 deg. Fahrenheit. Backyard stands of sugar cane are possible in colder climates, however. Cultivation styles vary considerably, but in general extensive, flat, low-lying fields are utilized and these are plowed deeply. Cuttings of varying length made from the upper joints of old canes propagate the sugar cane. These cuttings, known as seed, are placed in trenches and nearly covered with soil. They begin to sprout in about two weeks. When the cane is grown for human consumption, the cuttings are usually placed in holes. The crop has to be cultivated, weeded and fertilized extensively during the first few months. It is harvested from 10-20 months after sprouting. Harvest is 10-20 months after sprouting. The sugar content is carefully monitored and the canes are cut at just the right stage. This is usually when the flowers are beginning to fade.

The stems are cut close to the ground because the lower end of the cane is richest in sugar. Cane knives have been ordinarily used in poorer countries. The rhizomes normally give rise to two or three more crops, known as ratoons, before another planting is required. There have been up to 20 ratoon crops obtained, however.

Cultivated varieties today are usually hybrids of *Saccharum officinarum*, the "noble cane," with other hardier species.

Sometimes small owners of a stand of sugar cane extracted their own sugar in a primitive mill, but more often large "centrals" have been established which draw their supply from a wide area. In the milling process the canes are first carried to crushers where they are torn into small pieces. They are then passed through three sets of rollers. In the first set 2/3rds of the juice is expressed. They are then sprayed with water to dilute what sugar remains, and are passed through the second set.

These rollers exerts a very high pressure and remove nearly all of the moisture. After passing through the last set the residue is almost dry. This bagasse, as it has been named, can be used as a fuel for the mills, as a source of paper or wallboard because of its fibrous nature. It also contains a wax with some commercial value.

The juice that flows from the mill is a dark-grayish sweet liquid full of impurities. It contains sucrose, and other sugars, together with proteins, gums, acids, colouring materials, soil and pieces of cane. The purification of the sugar involves the separation of the insoluble materials and the precipitation of the soluble nonsugars. The juice is first strained or filtered to remove the solid particles.

It is then heated to coagulate the proteins, a process which is aided by the addition of sulfur. Lime is then added to neutralize the acids present, to prevent the conversion of sucrose into lower carbon sugars and to precipitate some of the substances in solution. These are removed by a series of filter bags or a filter press. Carbon dioxide may be added to aid in the process. The chemical processes involved in the purification of sugar are under constant supervision. The juice is now clear and dark coloured and ready for concentration. It is boiled down to a syrup of such density that the sugar crystallizes out. This operation is done in open kettles or vacuum pans.

The resulting sticky mass is known as massecuite. It is placed in hogsheads with perforated bottoms. The juice slowly percolates through the holes leaving the crystals of sugar behind. The juice constitutes the familiar molasses of commerce. In modern refineries the massecuite is centrifuged with the molasses passing out through fine perforations. The raw or crude sugar thus obtained is brown in colour and 96 percent pure.

Besides the bagasse, by-products of value are molasses, which is used in cooking and candy making. It is also used in the manufacture of rum and industrial alcohol. The better grades of molasses are obtained when the cruder methods of sugar milling are employed, for in such cases the sugar content of the molasses is higher. A mixture of bagasse and molasses, known as molascuit, is a valuable cattle feed.

Refining is the final stage of sugar preparation for markets. This is usually done in factories located in seacoast areas of the United States and Europe. The process involves washing to remove the film of dirt from around the crystals of crude sugar, dissolving the sugar in hot water, the removal of any mechanical impurities by filtering through cloth, decolouring by passing through bone black, recrystallization by boiling, and the removal of the liquids from the granulated sugar by centrifuging or other means. A hundred pounds of raw sugar usually yields 93 lb. of refined sugar and three-quarter gallon of refined molasses.

The granulated sugar is washed, dried, screened, and packed. Loaf, cube and domino sugars are made by treating granulated sugar with a warm concentrated sugar solution and pressing it into molds. Loafsugar is often sawed into blocks, strips or other forms. Powdered sugar is made from loafsugar or imperfect pieces of other types by grinding, bolting and mixing with starch to prevent lumping.

The refining of sugar is a very old process and was probably first done in North Africa. The first type of refined sugar was the sugar loaf, which appeared in England in 1310 and was familiar in America until late in the 19th Century. A marketable syrup has also been made from sugar cane by clarifying the juice and merely evaporating it to a consistency where the water content is 25-30 percent. This is sometimes called Golden Syrup.

India led the production of cane sugar in 1947, raising about 25 percent of the world's crop. Cuba, Brazil, Puerto Rico and other islands of the West Indies and Australia followed in the order names. In the United States Louisiana, Florida and Texas produce most of the sugar. The highest yields per acre have been recorded in Mauritius.

Sugar Beets: Sugar beet, *Beta vulgaris*, is another important source of sugar. It was derived from the wild *B. maritima*, which is still found wild on European seacoasts. There were times when beet sugar equaled or even exceeded that of cane sugar. However, by the 21st Century only about one-third as much beet sugar as cane sugar was being produced.

Although sugar beet was known since before the Christian era it was not used as a source of sugar until modern times. The leaves are edible as a substitute for spinach and the cooked beet serves as a delicious vegetable. The occurrence of sugar in the tubers was first noted in 1590 but Maregraf in 1747 first realized its possibilities. The industry formally began around 1800 in both Germany and France.

Napoleon promoted the use of it as an embargo against British importations. He was ridiculed because of this and a famous cartoon pictured him dipping a sugar beet into his coffee. Another cartoon showed him offering one to his small son, the King of Rome, with the caption, "Suck dear, suck, your father says it's sugar" (Hill 1952). Subsequently interest in sugar beets decreased, but it was again revived in France around 1829 and in Germany in 1935. Since then it has been a crop of increasing importance in many European countries. Although earlier attempts were made, the beet industry did not truly begin in the United States until 1879.

Sugar beet is a white-rooted biennial that grows best in regions where summer temperatures range around 70 deg. Fahrenheit. It produces well in irrigated portions of the southwestern United States during winter months. Almost any good soil is suitable for this plant's growth. Plants are frown from seed and must be thinned until they are from 8-10 in. apart. Thorough weeding and deep cultivation are required.

The crop is easily machine cultivated and harvesting, so is less expensive to raise than sugar cane. In temperate climates the seeds are sown inn April and the roots are allowed to remain in the ground until October because the sugar content increases in that position. The beets are pulled before the ground hardens and the tops are removed to prevent loss of sugar after which they are stored. The finest plants are saved for seed, which has led to a gradual improvement, by selection.

Extraction of the juice is a simpler process than for sugar cane because the roots are soft and pulpy. Earlier they were rasped to a pulp and the juice squeezed out in bags, but then a diffusion process was used. The roots are cleaned, cut into thin strips and heated in running water in a series of tanks. About 97 percent of the sugar can be extracted in this manner.

The waste beet pulp is removed and a process known as carbonation precipitates the insoluble impurities in the raw juice out. For this the raw juice is treated with lime, which coagulates some of the nonsugars, and carbon dioxide, which precipitates calcium carbonate. This settles out along with the impurities and the purified juice is separated out by filtration.

The process is repeated several times during which sulfur dioxide is added to adjust the alkalinity. A clear liquid is left after filtration, which is concentrated, crystallized and centrifuged just as with cane sugar. The massecuite is reboiled several times. It is difficult to differentiate between raw beet sugar and raw cane sugar for they are for all practical purposes identical in composition and appearance.

By-products of the industry include the green tops, which are used for cattle feed and fertilizers; the wet or dried pulp, which is a valuable cattle and sheep feed; the filter cake, which is used as a manure; and the molasses, which is used for stock feeding or for industrial alcohol.

Central Europe has produced most of the beet sugar, while in the United States California, Colourado and Idaho are the principal states that grow this crop. It might be noted that in the manufacture of rum in Europe, the flavour when the rum is derived from beet sugar differs substantially from that prepared from sugar cane. The former tends to be much more aromatic and has a distinctly unique taste

Maple Sugar: The sap of maple trees is used in making syrup and sugar. The area is confined to northeastern North America and was discovered and developed by the Amerindians.

Although several species of maple have a sweet sap, the most important is the sugar maple, *Acer saccharum*, and the black maple, *A. nigrum*. The sugar maple is a prominent tree in the northern part of the eastern deciduous forest zone. It reproduces naturally and lives to an age of 300-400 years. The red and silver maples give a comparatively small yield and are not of commercial interest. The sap begins to flow about the middle of March and continues for about a month. This is a period of warm, sunny days and cold nights. The best flow comes when the temperature reaches 25 deg. Fahrenheit at night and 55 deg. Fahrenheit during the day. The preferred location for tapping is the first three inches of sapwood, about 4.5 ft. above the ground.

Amerindians made incisions in the bark or large roots and conveyed the sap by reeds or curved pieces of bark into clay or bark receptacles. They boiled down the sap by dropping hot stones into it and converted the sap into sugar by letting it freeze and skimming off the ice. European settlers were quick to adopt the procedure and they added many improvements, eventually tapping with a 1-inch auger and using spiles to convey the sap into containers. They evaporated the sap in the open in large kettles so the sugar had many impurities.

The dark-brown sap was stored and later converted into sugar. This involved the famous sugaring-off, a process in which the syrup was boiled until it became waxy and the container was then dropped into snow. It was then poured into molds, where it immediately crystallized. Such simple methods are persisted for domestic use.

Commercial production further advanced the technique. Modern evaporators have replaced the furnaces and oiling pans of earlier times. These are able to convert from 25-400 gallons of sap into syrup in one hour. Great improvements have also been made in cleanliness and in the methods of collecting and transporting the sap. Such large operations involve from 100-1,000 trees. A well-managed sugarbush has about 70 trees to the acre, which allows room for the development of individual trees.

The maple-sugar industry reached its peak in 1869 when 45-million pounds were produced. With the advent of cane sugar it ceased to be an important commodity. Today the product is very pure and the demand is increasing for the syrup especially. Southeastern Canada is the leading producer while in the United States Vermont, New York, Ohio, Michigan and Wisconsin produce small amounts.

Palm Sugar: Several species of palm provide a fourth source of commercial sugar all of which is only available in the tropics and subtropics. The species utilized are the wild date, *Phoenix sylvestris*, the palmyra palm, *Borassus flabellifer*, the coconut, *Cocos nucifera*, the toddy palm, *Caryota urens*, and the gomuti palm, *Arenya pinnata*. Some of the oil palms also yield sugar. The date palm is tapped similar to that of a maple and the sap is obtained from the tender upper portion of the stem. In the other palm species the sap is secured from the unopened influorescences.

Usually the tip of these is cut off and the sap oozes out and is collected in various containers. The yield of this sweet juice, known as toddy, amounts to 3-4 quarts per day for a period of several months. The sap has a sugar content of about 14 percent. It is boiled down to a syrupy consistency and poured into leaves to cool and then hardens into the crude sugar known as jaggary. Some of this has reached European markets. Three quarts of juice yield about one pound of sugar. The toddy is often fermented to make the beverage known as arrack. The palm sugar industry is very old in India where over 100,000 tons were still being produced yearly by the mid 1900's.

Sorghum Syrup: The stem of the sweet sorghum, *Sorghum vulgare var. saccharatum*, contains a juice that is used in making syrup. To differentiate between a true syrup and a molasses it is necessary to realize that syrup is the product obtained by merely evaporating the original plant juice so that all the sugar is present. On the other hand, molasses is the residue left after a juice has been concentrated to a point where much of the sugar has crystallized out and been removed. The sweet sorghum or sorgo is a wild plant of the tropics and subtropics which ahs long been cultivated in many parts of the world.

The juice is a poor source of sugar, but yields nutritious and wholesome syrup. The stems are easily crushed and the juice is evaporated in shallow pans. On the average about 11-million gallons were being made in the United States by the mid 1950's. Similar syrup has also been made from sugar cane by clarifying the juice and evaporating it to a consistency where the water content is 25-30 percent.

Glucose Sugar: This sugar is also known as dextrose or grape sugar. It is present in small amounts in many of the organs of higher plants and is especially characteristic of fruits. However, commercial glucose is prepared from starch.

Fructose: Also known as levulose, this fruit sugar is present in many fruits together with glucose. It is somewhat sweeter than cane sugar and is valuable because diabetics can consume it. Commercial fructose is prepared from inulin, a polysaccharide that occurs in the tubers of the dahlia, *Dahlia pinnata*, the Jerusalem artichoke (= sunchoke), *Helianthus tuberosus*, and some other plants

Mannose: Mannose does not occur free in nature so that it must be obtained by hydrolysis from several complex compounds. It is readily oxidized from the juice of the manna ash, *Fraxinus ornus*, a tree in Sicily and southern Europe. The juice oozes from slits made in the bark and dries into a very sweet flakelike material known as manna. Its use is primarily in medicine.

Maltose: Maltose is also rarely found in a free state in plants, but is easily produced from starch through the activity of the enzyme diastase. It is used primarily in the brewing industry. Maltose syrup is sometimes used as a substitute for glucose and in medicine. Maltose that is obtained from rice starch has been used in Japan as a flavouring for over 2,000 years.

Honey: Flowers that are attractive, such as roses, hibiscus, etc., usually produce a sweet substance called nectar. This serves to attract various insects that are necessary for pollination of the plant. Nectar is composed mainly of sucrose with some fructose and glucose. It is used as food by bees, and some of it, after partial digestion, is converted into honey and stored for future use. During this process the sucrose is changed to an invert sugar, which is a mixture of fructose and glucose. Honey contains 70-75 percent invert sugar along with proteins, mineral salts and water. The sugar has a tendency to crystallize out. Honey was most likely the first sweetening substance used by humans. Beekeeping is one of the very oldest industries.

The flavour and quality of honey vary depending on the source of the nectar. Flowers that contain essential oils impart a typical taste. The bees favour certain plants and these are often cultivated near the apiaries. Clover, Alfalfa, buckwheat, lindens and some of the mints and citrus are among the favourites. Honey is an excellent food for it is almost pure sugar. It is also used in medicine, in the tobacco industry and in the manufacture of a fermented beverage called mead.

Starches and Starch Products

One of the most important and widely available vegetable products, starch constitutes the principal type of reserve food for green plants. It is a complex carbohydrate. It is stored in thin-walled cells in the form of grains. There are several types of starch that differ in the size and shape of the grains and other physical and microscopic characteristics. Important sources of starch are the cereal grains and underground tubers, although nuts, legumes and other plant organs may contain substantial amounts. Besides being a staple food for animals and humans, it also has many industrial applications.

Soluble starch is a form that is used in the textile industry to strengthen the fibres and cement the loose ends thus making a thread that is smoother and easier to weave. It also gives a finish to the end product. It serves as a mordant in calico printing, a thickener for the colours. Starch is also used as a sizing agent in paper industry, in laundry work, in medicine, in the preparation of toilet powders, as a binding for china clay and as a source of many other products.

Sources of Commercial Starch

Relatively few plants are used for the commercial production of starch. The main ones are potato, maize, wheat, rice, cassava, arrowroot and sago.

Cornstarch

Maize or Indian corn is the source of over 80 percent of the starch that is made in the United States. The grains are soaked in warm water with a small amount of sulfurous acid to loosen the intercellular tissue and prevent fermentation. Then the corn is ground so as not to injure the embryos. The ground material is placed in germ separators where the embryos are removed. The starch material is then ground very fine and is either passed through sieves of bolting cloth or is washed in perforated cylinders to remove the bran. The resulting milky liquid is run onto slightly inclined tables where the starch grains settle out and the remaining material flows off. The starch is later collected and dried in kilns and is then ready for the market. The best grades of cornstarch are used for food while inferior grades are for laundry starch and sizing and as a basis for glucose.

Potato Starch

Cull potatoes are utilized for making starch. These are washed and reduced to a pulp in graters or rasping machines. The resulting paste is passed through sieves to remove fibrous matter. After washing the solid starch is separated by sedimentation, the use of inclined tables, or centrifuging, and is then dried. Potato starch finds uses in the textile industry and as a source of glucose, dextrin and industrial alcohol. Europe is the principal producer.

Wheat Starch

The oldest commercial sources of starch were from wheat. It was known to the Greeks and was widely used in Europe in the 16th Century in connection with the linen industry. The gluten in wheat makes the removal of the starch a difficult process. It is accomplished by extraction with water or by the partial fermentation of the grain. Wheat starch is used mostly in the textile industry.

Rice Starch

Rice grains that are broken or imperfect are used for making rice starch. These are softened by treating with caustic soda and are then washed, ground and passed through fine sieves. More alkali is added and after a time the starch settles out as a sediment. This is removed, washed and dried. Occasionally dilute hydrochloric acid is used to free the grains. Rice starch has found use in laundry and for sizing.

Cassava Starch

Cassava flour and tapioca are used in industry mainly as sizing materials and as the source of certain starch products.

Arrowroot Starch

The tubers of several tropical plants provide a source for arrowroot starch. West Indian arrowroot is from *Maranta arundinacea*. Florida arrowroot is from *Zamia floridana*. Queensland arrowroot is from *Canna edulis*, and East Indian arrowroot from *Curcuma angustifolia*. The tubers of these plants are peeled, washed and crushed and the pulp passed through perforated cylinders. A stream of water carries the starch into tanks where it settles out. Arrowroot starch is easily digested and thus is valuable as a food for children and invalids. There is little use in industry.

Sago Starch

The stems of the sago palm, *Metoxylon sagu*, contain starch. Cultivation is in Malaya and Indonesia. The flowers appear when the trees are about 15 years old and just prior to this time the stems store up a large amount of starch. The trees are felled and the starch pith is removed. This is ground up, mixed with water and strained through a coarse sieve. The starch is freed from the water by sedimentation and washed and dried. This is known as sago flour. Commercial sago is prepared from this by making a paste and rubbing it through a sieve in order to cause granulation. The product is dried in the sun or in ovens and appears as hard shiny grains, known as pearl sago. Both sago starch and pearl sago are used almost entirely for human consumption.

Starch Products

Soluble Starch

Starch grains are insoluble in cold water but they readily swell in hot water until they burst to form a thin, almost clear solution or paste. This soluble starch has been used for finishing textiles and in the paper industry.

Dextrin

When starch is heated or treated with dilute acids or enzymes it becomes converted into a tasteless, white, amorphous solid known as dextrin or British Gum. Dextrin possesses adhesive properties and has been used as substitutes for mucilage, glue and natural gums. Bread loaves brushed with dextrin aids in crust formation. In steel manufacture, the sand for the cores used in casting is held together with dextrin. Other uses include cloth printing, glazing cards and paper and making pasteboard.

Glucose

When starch is treated with dilute acids for a long time it becomes more completely hydrolyzed and is converted into glucose sugar. Often the same factory that extracts the starch also converts it into glucose. This operation is done in large copper boilers under pressure. About six pounds of dilute hydrochloric or sulfuric acid are used for each 10,000 pounds of starch. After all the starch has been converted, the free acids are neutralized with caustic soda. The liquid is then decolourized with boneblack and concentrated into thick syrup. One of the common brands of corn syrup is "Karo." Glucose may be considered as an inferior substitute for cane sugar. However, it is an excellent food material. Its use is in table syrup, for sweetening, in candies, jellies and other kinds of cooking. It is often mixed with maple syrup, brown sugar, honey or molasses it is used for making vinegar and in brewing.

Industrial Alcohol

Starch is the source of an enormous quantity of industrial alcohol. Maize and potatoes constitute the chief sources, although the other starches and even cellulose, various products of the sugar industry and fruit juice may be utilized. The process converts the starch into sugar by means of diastase and the fermentation of the sugar by yeasts to yield alcohol. The operations are carried out under different conditions from those followed in making alcoholic beverages. When fermentation has stopped, the alcohol is extracted from the mash by fractional distillation. The alcohol thus formed as a result of the fermentation of sugar is known as ethyl alcohol, as distinguished from methyl or wood alcohol, a product of the destructive distillation of wood. To render it unpalatable, ethyl alcohol is often "denatured:" by adding methyl alcohol or other substances. Industrial alcohol is the most important and most widely used solvent and is the basic material in the manufacture of hundreds of products. It is also used in medicine, pharmacy and other industries.

Nitrostarch

Starch and cellulose are chemically very similar products. Cellulose reacts with nitric acid to form nitrocellulose while starch yields nitro starch. Nitrostarch is a very safe explosive if the ingredients are absolutely pure. Tapioca starch was originally imported for this purpose but during World War I cornstarch was used as a source.

Cellulose Products

The most complex of the carbohydrates, cellulose is present in the cell walls of all plants. Because of their strength, cells with thick walls have been used in various industries. Besides the natural product being used in the textile industry, artificial fibres are derived directly from cellulose as well as countless other products. Cellulose chemistry is an important phase of organic chemistry.

Cotton, a very pure form of cellulose, has been used for a very long time in the production of artificial fibres and other cellulose products. Wood is another very available source. When certain woods are treated with concentrated acids or alkalis, the bond between the wood fibres and the lignin, which cements them together, is broken, and the fibres, which are pure cellulose, can be removed. These fibres may then be reorganized as paper, or they may be treated further chemically. If the chemical treatment merely causes the dissolution of the fibre into its component molecules, these molecules may be synthesized into artificial fibres or converted into cellulose plastics. But if the molecules themselves are broken down, their component elements, carbon, hydrogen and oxygen, may be recombined to form wood sugar. Thereafter the wood sugar may be transformed into yeast or alcohol and thus become available for food or as the raw material for numerous industrial products.

Paper and Paper Industry

A very important use of cellulose is in the manufacture of paper, a very old industry. The word "paper" comes from the Latin "papyrus" the name of a sedge, the pith of which was used for paper in 2400 B.C. Egypt. However paper was first made in China. The industry spread from China to India, Persia and Arabia and then through Spain to other European countries. The first paper mill in the United States was in 1690 at Philadelphia.

Raw Materials

The papermaking value of the various fibres depends on the amount, nature, softness and pliability of the cellulose present in the cell walls. This cellulose may occur alone or in combination with lignin or pectin. Wood fibres, cotton and linen are the principal raw materials.

Wood Fibres

Wood began being used in the paper industry from about 1850. Today wood has largely replaced the other fibres and furnishes over 90 percent of all the paper manufactured in the Untied States.

Spruce is a very important source of wood pulp and has furnished about 30 percent of the total supply. It is ideal because it has all the requirements of a good pulpwood. The fibres are long and strong with a maximum content of cellulose. The wood is almost free from resins, gums and tannins; and it is light coloured, sound and usually free from defects. Red spruce, *Picea rubens*, white spruce, *P. glauca* and Sitka spruce, *P. sitchensis*, are the main species.

The southern yellow pine, *Pinus australis*, is another important pulpwood. The eastern hemlock, *Tsuga canadensis*, is a main species in the Lake States while the western hemlock, *T. heterophylla* is important on the Pacific Coast. Other species include aspen, *Populus grandidentata* and *P. tremuloides*, and balsam fir, *Abies balsamea*. Of lesser importance are jack pine, *Pinus banksiana*, tamarack, *Larix laricina*, white fir, *Abies concolour*, and several hardwoods among which are the beech, *Fagus grandifolia*, sugar maple, *Acer saccharum*, and birch, *Betula lutea*. Sawmill waste is also an increasingly valuable source of wood pulp.

Cotton and Linen

Up to the middle of the 19th Century rags of cotton and linen were the only source of paper, and they are still used for making the finest grades. Cotton fibres have a high felting power and a high cellulose content of about 91 percent. Rags and raw cotton in the form of fuzz or linters are utilized. Flax fibres, that comprise linen, contain 82 percent of pectocellulose and yield a paper of great strength, closeness of texture and durability. Textile waste may also be used. In the preparation of rag pulp the material is sorted, cut into small pieces and freed from dust. It is then boiled in caustic soda to remove the grease, dyes and other impurities and it is washed until perfectly clean. The resulting pulp is used directly in papermaking.

Minor Paper Raw Materials

Esparto

Esparto grass, *Stipa tenacissima*, is important in Great Britain. The plant is native to Northern Africa where it flourishes in dry, sandy and rocky coastal regions. The tufted wiry stems are plucked and pressed into bales for shipment. Esparto is converted into pulp by heating in a caustic coda solution under pressure. Although the cellulose content is only 48 percent, the fibre has great flexibility and felting power. It yields an opaque, soft, light paper of uniform grade. The finest printing papers in England have been made from esparto, either entirely or in a mixture. Another grass, *Lygeum spartum*, is also used.

Textile Fibres

There are other textile fibres other than cotton and flax that have been used as sources for paper. The waste material of the jute and hemp industry, in the form of old ropes, sacking, sailcloth, etc. was once used extensively for making strong and tough papers. Jute and jute butts were used mainly for wrapping paper, envelopes, cable insulation, while hemp, after bleaching, yielded ledger and bank-note paper. Manila hemp was an important source of envelopes and wrapping paper. Ramie, sisal, sunn hemp, New Zealand hemp, coir and other fibres have been utilized.

Paper Mulberry

The bast fibres of the paper mulberry are very strong and have been used in Japan for paper lanterns and umbrellas and as paper for writing. The fibre is scraped, soaked and beaten, after which it is mixed with mucilage and spread on frames to dry. When treated with oil this paper becomes strong enough to serve as a substitute for leather or cloth.

Misc. Paper Sources

There are many other sources for making paper than what has been noted previously. For example, stems of rye, barley, wheat, oats, rice and other grasses are used for low-grade paper, strawboard, pasteboard and cardboard. These plants have fibres with low cellulose content and are too short and small to have much tensile strength. Thus, they are mixed with other fibres. Sugarcane bagasse, cornstalks, and waste paper have been developed as sources of paper. Banana fibre, tree bark, rushes, weeds, broomroot, licorice, cotton and tobacco stalks, beet pulp waste and peat have been used somewhat. Recycled paper is extensively used to make other paper products.

In the Orient bamboo fibre is an important source of paper. The papyrus, *Cyperus papyrus*, baobab, *Adansonia digitata*, and *Daphne cannobina* are used in Africa and India. Rice paper of Japan and China is made principally from *Tetrapanax papyriferum*, *Edgeworthia tomentosa* or *Wickstroemia canescens*.

Synthetic Fibres From Plant Products

Hill (1952) writes that since the Middle Ages until modern time there have been many schemes to make artificial silk and other fabrics. In 1880 Count de Chardonnet made the first artificial fibre, and a few years later the first artificial silk. Later factories were established for making the product. At the outset this new material was handicapped by its name for the public considered it only as an imitation or a substitute. This condition existed as late as 1908. Today it is understood that all the artificial fibres constitute entirely new products with valuable characteristics and properties of their own.

The raw material of the rayon industry is high-polymer alpha cellulose, prepared in a pure form from wood pulp or cotton linters. Purification is accomplished by the elimination of mechanical impurities through air separation and then cooking in a 3.5 percent sodium hydroxide solution. This removes all the other organic substances.

The pure cellulose fibres that remain are bleached, washed and dried. The next step is to dissolve the cellulose by various solvents thus rendering it sufficiently liquid so that it can be squirted in a fine jet. The solution is then forced by pressure through minute perforations in glass or platinum, and emerges from these "spinnerets" in thin streams. The streams are coagulated into fine, almost invisible filaments in different ways. The solvents are removed and the filaments are caught up by revolving reels and twisted into threads suitable for spinning. The threads are washed, bleached and dried. In the viscose, nitrocellulosee, and euprammonium processes the final product is an almost pure cellulose fibre known as regenerated cellulose. This is chemically identical with the cellulose in cotton, but it differs in its mechanical properties. The product of the acetate process is a cellulose ester , cellulose acetate. This differs from regenerated cellulose in both its chemical and physical properties. Several kinds of rayon are Viscose Rayon, Cellulose Acetate Rayon, Cuprammonium Rayon and Nitrocellulose Rayon.

Cellulose Nitrate Products

Treatment with concentrated nitric acid in the presence of sulfuric acid causes cellulose to change into several types of cellulose nitrate. These differ according to which concentration of nitric acid was used and the consequent degree of nitration as well as the temperature and the duration of the action. The higher cellulose nitrates are called guncotton, or in error nitrocellulose. The lower nitrates constitute pyroxylin, or collodion cotton.

Guncotton

This is made from cotton linters during which process the cellulose is completely nitrated. It is used as an ingredient of many high explosives. Cordite, e.g., is a combination of guncottton and nitroglycerin, while smokeless powder is made from a mixture of guncotton and the lower nitrates. Guncotton is one of the safest of explosives to handle when properly manufactured.

Pyroxylin

A partial nitration of cellulose produces pyroxylin. This is carried out under different conditions from those, which result in the formation of guncotton. In modern photography the films often consist of pyroxylin coated with gelatin. It is used in the rayon industry. However, its chief value is that it is soluble in a variety of solvents and yields many useful products, such as celluloid and other plastics, collodion, artificial fabrics and varnishes.

Cellulose Acetate Products

Industrial uses include as a substitute for cellulose nitrate in the film industry because it is not very flammable. Resultant films are more brittle and expensive, however. It is also used for goggles, gas masks, automobile windows, artificial fabrics, index cards, airplane varnish, etc.

Viscose Products

Cellophane is a viscose product. Forcing crude viscose through tiny slits rather than perforations makes this. It coagulates into a thin transparent film only one-thousandths of an inch in thickness. Such viscose files are used for numerous poses even for sausage casings. Viscose fibres have replaced cotton in Welsbach mantles.

Products of Cellulose Hydrolysis

When cellulose is completely hydrolyzed in a process of saccharification this ultimately results in the conversion of the cellulose into wood sugar that in its turn yields alcohol and yeast. It is a complement to the manufacture of paper pulp from wood.

Under usual processes of pulp making one ton of wood is converted to one-half ton of pulp and enough wood sugar to yield 10-12 gallons of alcohol.

Doubling the cooking time produces an edible starchy material that is an excellent cattle feed, and sufficient sugar to furnish 20 gal. of alcohol is obtained. Under a still longer cooking time all of the cellulose can be converted into sugar with an eventual production of 60-80 gal. of alcohol.

Regarding the kinds of wood sugar that are produced, conifers yield glucose, pentose, and mannose; hardwoods glucose and pentose. These sugars are useful because they can be readily fermented.

Glucose and mannose, which comprise about 2/3rds of the wood sugar, are converted into ethyl alcohol by the action of ordinary yeast. This industrial alcohol has a wide range of uses. Pentose sugars are not fermentable into alcohol, but they can be converted, through the activity of a bacterium, *Torulopsis utilis*, into Torula yeast, and edible substance with a 50 percent protein content. One ton of pulp will yield from 40-100 lbs. of yeast.

Cellulose hydrolysis can play an increasing role in a well-organized system of forest utilization as it eliminates much of the waste. Small pieces of wood, chips, sawdust, wood flour and sawmill waste can all be used effectively

Hemicellulose

Many tropical plants have seeds that are very thick, hard and have heavy walls consisting of hemicellulose. This is a modification of ordinary cellulose and constitutes a supply of reserve food for the plant. In young seeds the endosperm consists of a milky juice, but as the seeds mature this fluid is replaced by the harder material. hemicellulose cannot be used by animals as food. However, it does provide vegetable ivory of commerce.

Vegetable Ivory

The Ivory-nut or Tagua Palm, *Phytelephas macrocarpa*, of the Neotropics is the main source of vegetable ivory. The palm is a low-growing tree typically on riverbanks from Panama to Peru. The drupes like fruits have from 6-9 bony seeds with a thin brown layer on the outside and a very hard and durable endosperm.

The natives collect the seeds that are shipped to Europe and the United States. Ecuador is the main exporter. This ivory may be carved and shaped into various objects, so it serves as a substitute for true ivory in the manufacture of buttons, poker chips, knobs, chessmen, dice, inlays, etc.

Several species of palms in Africa and Polynesia and Neotropical America have seeds of similar constituency. With the exception of *Metroxylon amicarum* of the Caroline Islands, these are not important in commerce.

Tanning, Dye & Processing Materials

Tanning

Tannins are organic compounds, mostly glucosidal, which have an acid reaction and are very astringent. Their biological function is in doubt. They may be concerned with the formation of cork or pigments, or with the protection of the plant. Tannins are of economic interest because of their ability to unite with certain types of proteins, such as those in animal skins, to form a strong, flexible, resistant insoluble substance known as leather. Because of this characteristic, tannin-containing materials are in great demand. Tannins also react with salts of iron to form dark-blue or greenish-black compounds, the basis of common inks. Their astringent nature makes them useful in medicine. Tanning materials are also utilized in oil drilling to reduce the viscosity of the drill without reducing the specific gravity. Most plants contain some tannin, but only a few species have a sufficient quantity to be of commercial importance. Tannins are found in the cell sap or in other definite areas in bark, wood, leaves, roots, fruits and galls. Such structures are of little value for other purposes, so that the extraction of tannin is usually incidental to other industries.

Tanning Industry

Hill (1952) described the tanning industry as very old, possibly dating to 3,000 B.C. Leather was being tanned in China before 1,000 B.C. The Romans used oak bark for tanning animal hides and the Amerindians used several native plants to cure the hides of buffalo. In the United States the first tannery was established in Virginia in 1630. Scarcely 20 years after there were more than 50 tanneries in New England. The industry centred around the latter because of the abundant hemlock bark available there. Thereafter the industry gradually shifted to the west and south when hemlock supplies became scarce. Oak then became the principal tanning substance. Later on chestnut was used in the South. When these became scarce other sources, such as sumac and canaigre, were utilized and some foreign products were imported. Today concentrated extracts with high tannin content are generally utilized.

Sources of Tanning Substances

Oak

Several species of American oaks used for tanning. The Chestnut Oak, *Quercus montana,* is abundant in Appalachia. It has tannin content of 6-11 percent. The extract with 26-30 percent tannin is widely used for heavy leathers. The Black Oak, *Quercus velutina,* produces quercitron extract. Although widely used it does stain the leather with a yellowish colour. The California Tanbark Oak, *Lithocarpus densiflora,* was utilized since 1850. It has tannin content of as high as 29 percent. Red Oak, Quercus borealis, and White Oak, Q. alba, are used to some extent but they have lower amounts of tannin. European oaks are used in England and on continental Europe.

Mangrove

Mangrove is a good source of tannin and gradually became more important as other sources diminished. The Red Mangrove, *Rhizophora mangle,* has been the main source. It is abundant in tropical swamps of both hemispheres. The bark is very hard and heavy and contains 22-33 percent tannin. The leaves may also be used and the extract has been the least expensive of tanning substances. However, it causes the leather to darken and is therefore rarely used alone.

Wattle

This source is used mainly in Great Britain. It is obtained from several species of acacia, mainly *Acacia decurrens* and its varieties *dealbata* and *mollis*, and *Acacia pycnantha*. These are small trees native to Australia but have become cultivated in Africa, Ceylon, Brazil and other countries. The bark may have tannin content of 40-50 percent. It is removed from trees that are 5-15 years old. It is then ground to a powder. Wattle bark produces a solid, very firm, and faintly pink leather, which is especially used for soles. Wattle wood is a good pulpwood and can be used for timbers, posts and other things. There is also a gum produced by the trees.

Other Bark Sources

Other barks that are occasionally used for tanning are the European larch, *Larix decidua*, Norway Spruce, *Picea abies* and some birches and willows. Birch bark is favoured in Russia and the fragrance of Russian leather is due to the presence of an essential oil in the bark. Willow bark gives light-coloured, soft pliable leather favoured for tanning gloves. Some tropical barks are of minor importance. Mallet bark is obtained from *Eucalyptus occidentalis* of Western Australia, with tannin content of 35-50 percent. Avaram bark is from *Cassia auriculata*, important in India. Tanekaha bark from *Phyllocladus trichomanoides*, a New Zealand tree is used for gloves because it contains an orange-yellow dye.

Chestnut Wood

Castanea dentata and *Castanea sativa* contain tannin that is extracted at high temperatures from wood chips. The final product is 30-40 percent tannin. Chestnut is used for heavy leathers.

Quebracho Wood

This wood of several South American trees is a very important source of tannin. *Schinopsis lorentzii* and *S. balansae* are the main species. The wood from these trees is some of the very hardest known with a specific gravity of 1.3-1.4. Argentina and Paraguay are the main producers. Extracts are made in factories located near the sources. The logs are chipped and cooked with steam in copper extractors until the liquor is very concentrated. This has tannin content of 40-60 percent. It is very quick in its tanning action and is used either by itself or in combination for all kinds of leather, especially sole leather because it imparts extra durable qualities.

American Sumac Leaves

The dried leaves of three North American species of sumac, *Rhus glabra, R. typhina* and *R. copallina,* are important tannin sources. Leaves are harvested in autumn when they begin to turn red, and are dried and ground to a powder. The tannin content of 10-25 percent is higher in plants that grow at lower latitudes.

Sicilian Sumac Leaves

Rhus coriaria has a content of 20-35 percent The leather becomes a pale colour and is soft in texture. It is especially suited for gloves and bookbinding. This is one of the few tanning plants that are cultivated.

Gambier Leaves

Gambier or White Cutch is a resinous substance that is extracted from the leaves and young branches of *Uncaria gambir*. It is a climbing shrub of Malaya and Indonesia. The plants become shrubby when cultivated. The trees are cropped four times a year and the tannin is extracted from the tissues with boiling water. It crystallizes out as a semisolid white substance and has tannin content of 35-40 percent. Gambier is also used as a dye, masticatory and in medicine.

Myrobalan Fruit

Myrobalan nuts are the unripe fruits of two trees of India, *Terminalia chebula* and *T. bellerica*. These trees are grown in India for both fruit and timber. The nuts have tannin content of 30-40 percent. When they are used by alone they yield spongy, light yellow leather, but in combination they are preferred. They are used to tan leather of goats, calves, and sheep and are best suited for soles and harnesses.

Divi-Divi Fruit

The tannin is obtained from the dried twisted seedpods of a small leguminous tree, *Caesalpinia coriaria,* of the West Indies and South America. The tannin content is 40-50 percent.

Tara Fruit

A stocky shrub or small tree Tara, *Caesalpinia spinosa* is widely distributed in tropical America. The fruits are 43-51 percent tannin. The plant is cultivated in Peru and North Africa. It is used for producing a high grade of leather as it changes the colour only slightly. The fruits are also used for ink and black dye.

Algarobilla Fruit

Caesalpinia brevifolia from Chile has pods with high tannin content. It is usually used in combination with other materials.

Valonia Acorn Cups

The sun-dried cups of Turkish oak acorns, *Quercus macrolepis,* have tannin content of 45 percent. It is used in combination with other materials for fine grades of leather.

Canaigre Roots

Rumex hymenosepalus is a species native to the southern United States and Mexico. It is widely cultivated. The roots are sliced and the tannin extracted. The tannin content is about 30 percent and yields a bright orange firm and heavy leather.

Palmetto Roots

Sabal palmetto has been used somewhat as a source of tannin, but the content is only 10 percent.

Ink Manufacture

Writing inks date back to ancient times. In Egypt ink was used on papyrus before 2,500 B.C. and the oldest writings in China are dated to at least 2,600 B.C. Originally carbon ink was used, which is a combination of gums, charcoal and varnish. The charcoal was secured from plant sources such as charred date seeds or it was of animal origin. The two most important modern inks are Carbon Inks and Tannin Inks.

Carbon Inks

These are paint like inks that remain on the surface of paper while the others are dyelike and soak into the paper where they combine chemically with the fibres. India or Chinese Ink is very permanent ink that is made from the carbon black, lampblack, or soot obtained by burning pinewood or a vegetable oil such as tung or sesame, mixed with glue, gum arabic, or some similar sizing material. Printing Ink contains carbon obtained from natural gas, petroleum, or other materials, combined with rosin, a drying oil such as linseed, some chemical drier and often soap.

Tannin Inks

These utilize the property of tannin that combines with iron salts to produce a blue-black colour. Most are gallotannate in nature, the tannin being derived from insect galls that also contain gallic acid. Tannin inks were first used in the 11th Century.

Aleppo or nutgalls have been the main source of the tannin. These galls are formed on the twigs of the Aleppo oak, *Quercus infectoria*, as a response to the injuries caused by egg laying insects. The plant is a small shrub that ranges throughout the Mediterranean region.

The small spherical or pear-shaped galls form in great quantity and have very high tannin content. To make the ink either the galls or an extract of them are combined with ferrous sulfate, an agglutinant like gum arabic, and a colouring material, like logwood. *Rhus chinensis* and other species of sumac from China and Japan are at times used to substitute the Aleppo galls, although they are not of as good a quality. Oak galls are sometimes used and tannin inks may also be made from other sources of tannin such as logwood and chestnut. Logwood has the advantage of containing both tannin and a colouring agent.

Coloured Inks: These are made from natural or aniline dyes in combination with water, gum and alum. A high quality red ink is derived from brazilwood.

Dyes and Pigments

Natural dyestuffs and stains, obtained from the roots, bark, leaves, fruit or wood of plants, have been in use worldwide from earliest time. The cultivation of the plants and the preparation of the dyes have been an important industry in many areas. Around the mid 1800's the natural products began to be supplanted by synthetic or aniline dyes obtained from coal-tar products. These synthetic dyes are brighter, more permanent, easier to use, are less costly and afford a wider range of colours. Their development has gradually led to the abandonment of most of the plant products.

There are over 2000 different pigments secreted by plants. Most of these are used only locally by primitive cultures, if at all. A comparatively small number, about 155, have been of commercial importance, and of these only a few have been able to compete with the artificial colours.

The principal use of dyes has been in the textile industry. However, before the fabrics can incorporate them, they must be rendered insoluble so that they will not run. This is accomplished by the use of mordants, which are various metal salts. When fabrics are steeped in a solution containing a weak salt of iron, chromium, aluminium, or tin, a fine layer of the metallic oxide is deposited on the cloth. The dye forms an insoluble compound with this oxide. Dyes are also used for colouring paints, varnishes, leather, ink, paper, wood, furs, food, cosmetics and medicines.

Many different kinds of dyes have been in use in the United States alone at various times. The Amerindians made use of many native species and the early settlers followed their example. In the past dyes that have been important include butternut bark, which was used for dyeing homespun, and later for dyeing the uniforms of the Confederate army.

As in other parts of the world plant pigments have been generally supplanted by aniline dyes. For a while after 1914 when the First World War reduced the supply of synthetic colours, 90 percent of which were made in Germany, the United States returned to the use of the natural products.

However, soon a domestic synthetic dye industry arose which by 1930 was producing products valued at US$83,000.000. Almost all colours were available that occurred in plant pigments. Red dyes were obtained from alkanna, barwood, brazilwood, cudbear, logwood, safflower, sappanwood and sandalwood. Yellow dyes were from annatto, fustic, gamboge, henna, osage orange, Persian berries, quercitron, saffron, tumeric and weld. Blues were from cudbear, indigo, and woad. Greens were from chlorophyll and loka9o; and brown was from cutch.

Wood Dyes

Logwood

This is one of the oldest and most important dyes. It is obtained from the heartwood of *Haematoxylon campechianum*, a small, thorny tree legume of Mexico. It has been introduced throughout the world tropics. The trees are propagated from seed when cultivated. They are cut when 10-12 years old and the bark and sapwood removed. Both the logs and/or the extracts are exported. The purple-red dye is known as haematoxylin and is used in its natural state or with a mordant. The presence of a large amount of tannin allows logwood extract to react with iron salts to produce a black colour. This is used especially for dyeing cotton and woolen fabrics, leather, silk and furs. Haematoxylin stain is widely used in histological work. A related species, braziletttte or hypernie (*H. brasiletto*) gives a red dye that is used for dyeing leather.

Fustic

This is the main source of natural yellows, olives and browns and ranks with logwood in importance. It is used for leather and in combination with logwood for silk, wool, nylon and rayon. It comes from the heartwood of *Chlorophora tinctoria*, a forest tree of the West Indies, Central and South America. The light-yellow wood turns a dark yellow-brown when exposed to air. Faustic is exported as short logs, chips, powder or paste. The dye is frequently called Old Fustic to distinguish it from Young Fustic, once obtained from the twigs of *Cotinus coggygria*.

Cutch

The term Cutch refers to several kinds of raw materials that are useful in making dyes and in tanning. Gambier or white cutch was discussed above. Black cutch or catechu is the source of an important brown dye. It is from the heartwood of *Acacia catechu*, a tree native to Burma and India. Wood pieces are boiled in water and the extract is evaporated to a purple-black, gummy, semisolid mass that is then molded into blocks. This is a fast dye and is used for the different shades of fawn, brown, olive and drab colours, including khaki. Catechu is also used in medicine and as a masticatory.

Osage Orange

Maclura pomifera is a tree native to southern Missouri and Texas. It is frequently planted as an ornamental. The bright orange wood yields a dye that is used for orange-yellows and gold and as a base for green colours. Amerindians used Osage Orange and it is a substitute for fustic and aniline dyes.

Sappanwood & Brazilwood

These are soluble red wood dyes that have had an interesting history. One of the first red dyewoods known was from the heartwood of *Caesalpinia sappan*, or sappanwood. The tree is indigenous in Malaya and India and cultivated elsewhere in Asia. The wood was introduced into Europe during the Middle Ages when it was called Bresel Wood. Then the Portuguese discovered a similar wood in South America to which they applied the name Bresel. The source of this Western Hemispheric Brazilwood is *Caesalpinia echinata*, and it ranges generally in the American tropics. The heartwood has a red dye that is used for woolen and cotton cloth and for red ink. The wood is also valuable for making violin bows. The bark and pods also serves as a source of the dye. Sappanwood has the same uses as Brazilwood, and in both the colour is fugitive.

Barwood & Camwood

These are obtained from several West African trees. *Baphia nitida* and a related species are usually the source of Camwood.

Barwood is obtained from *Pterocarpus erinaceus* and *P. soyauxii*. They both yield shades of red, brown, and violet and are used mainly for dyeing wool.

Red Sanderswood (Red Sandalwood): *Pterocarpus santalinus* is an East Indian tree with hard, fragrant, reddish wood that is the source of an insoluble blood-red dye.

Leaf Dyes

Indigo

Indigo or Anil was known as the "King of the Dyes," due to the permanency and strength of its deep-blue colour. Synthetics have largely replaced it today, however. It is obtained from *Indigofera tinctoria* of Asia and *Indigofera suffruticosa* of tropical America, as well as several other species of the same genus. These are stiff-stemmed, weedy annuals or shrubby perennials.

The dye is not present in the plant itself. The leaves contain a soluble colourless glucoside, indican, which oxidizes in water to form the insoluble indigo. Fresh plants are gathered in the flowering season, are broken up and steeped in water for 12 or more hours. The liquid is constantly stirred to bring about complete oxidation, and the indigo gradually settles out as a blue sediment that is packaged as small cubes for export. Indigo was used as a dye in India and other parts of Asia in ancient times. It was introduced into Europe in the 16th Century and from there it spread worldwide.

Chlorophyll

Being present in all green plants, chlorophyll is especially typical of the leaves of the higher plants from which it can be extracted with different solvents. Chlorophyll is important as a colouring substance for foods, soap and other products. It is especially valuable by being harmless and serving as a deodorant.

Henna

The leaves and young shoots of *Lawsonia inermis* provide an orange dye. It is a small tree 6-8 ft. tall and native to India, Iran and the Arabian Peninsula. The plant is widely cultivated in the tropics and subtropics as an ornamental and dye plant. Branches are cut when the tree is only three years old and from then on twice annually. The leaves are dried and ground to a paste. Henna is a fast dye that was once used mainly for leather and fabrics. It also serves for dyeing eyebrows, fingernails, hair and other personal adornment.

Woad

Isatis tinctoria is a blue dye that was used in Europe in early times. It was used by the primitive Britons to paint their bodies (Hill 1952). The leaves are moistened, slightly fermented, molded into balls and dried.

Weld

This is a deep yellow European dye that has been widely used for silks. It is extracted from the leaves of the weld, *Reseda luteola*, that was formerly cultivated in Europe. The American colonists brought the plant to America where it still persists in several localities.

Root and Tuber Dyes

Madder

Once widely cultivated in the Mediterranean region, Madder is still grown in the Levant and in Italy. The dye is found in the roots of *Rubia tinctorum*, which occurs as a wild plant in Asia Minor, the Caucasus and Greece.

An infusion is made from 3-4-year old roots to produce a brilliant scarlet colour known as "turkey red." The colouring agent is a glucosice, alizarin, which was one of the first synthetically produced dyes.

Alkanna: *Alkanna tinctoria* roots yield a red, violet or crimson dye that is used for colouring pomades, medicines, oils and wines and as a histological stain.

Turmeric: An important colouring material in India, turmeric is obtained from the tubers of *Curcuma longa*. The colour is orange-red or reddish brown. It is used to give a yellow colour to cloth and foods, such as curries. Turmeric also is a chemical indicator as it changes colour in the presences of acids or alkalis.

Bark Dyes

Quercitron: The crushed bark of black oak, *Quercus velutina*, of eastern North America yields a bright yellow dye called quercitron.

It is used to dye cotton, leather and wool. It is also a tanning substance. Flavin is a strong preparation of quercitron.

Lokao: Often called Chinese Green, this is one of the few natural green colouring substances. It is the powdered bark of two Chinese buckthorns, *Rhamnus globosa* and *R. utilis*. It use is in dyeing cotton and silks.

Flower Dyes

Safflower: *Carthamus tinctorius* is a dye plant native to India. The flowers are used to colour food and the seeds furnish an edible oil that is low in cholesterol. The leaves are also used in salads. The yellow or orange thistle like heads are picked in dry weather dried and pressed into cakes. An extracted red dye is used for fabrics and rouge and a yellow dye is for various purposes.

Saffron: *Crocus sativus* is the source of an old and strong yellow dye. Native from Asia Minor to Greece it is cultivated in many parts of Europe and Asia. The lavender coloured flowers blossom in autumn. The stigmas and tips of the styles contain the dye. These are clipped as soon as the flowers open and are dried naturally or by heating. At least 4,000 flowers are required to make one ounce of dye. The colouring substance is soluble in water so it may not be used for fabrics. However, it is useful for colouring medicines and food to which it also imparts a typical flavour.

Dyes from Fruits

Persian Berries: A buckthorn, *Rhamnus infectoria*, of Southern Europe, Asia Minor and Iran, has fruits called Persian Berries. An extract yields yellow and green dyes.

Sap Green: *Rhamnus cathartica* of Europe is also the source of Sap Green. It is a watercolour pigment obtained from the fruit. It also has medicinal properties.

Dyes from Seeds

Annatto: Seeds of *Bixa orellana* are the source of this dye. The plant is an evergreen shrub or small tree native to tropical America. Fruit is borne in the second year and the yield averages 300-600 lbs. per tree. Each spiny pod contains 30-50 seeds surrounded by a scarlet aril, the source of a bright yellow dye. Annatto is almost tasteless so that it is adapted for colouring cheese, butter, margarine and other foods. It is also sued to dye wool and calico, paint, varnish, lacquer and soap. In South America Amerindians paint their bodies red with Urucú, which is the same dye.

Gum Resin

Gamboge: The gamboge tree, *Garcinia hanburyi*, and a few other species, from Thailand exudes a gum resin that furnishes a yellow dye. Incisions are made in the bark that yields a yellow viscid fluid, which dries on exposure to air. The yellow dye is soluble in water, alcohol or oil and is widely used by artists. It also gives a gold tinge to varnishes that are used for lacquer and metalwork. Gamboge is a violent cathartic.

Dyes from Lichens

Archil and Cudbear: Several purple or blue dyes are known by various names such as Archil, Orchil, Orseille or Cudbear. They are obtained from different species of lichens, mainly *Roccella tinctoria.* These were used for wool and silks and for staining wood and colouring wine, but now they are used mainly for drugs in sauces and bitters. Treating macerated lichens with ammonia and exposing them to air make the dye. A blue archil liquor is then extracted with water. When heated to drive off the ammonia it changes to red archil. This is evaporated and ground to a fine power or paste known as cudbear.

Litmus is made from the same lichens by a different process. The lichens are treated with an alkali and allowed to ferment for a few days. Lime is added and the dye extracted with water. The liquid is evaporated and mixed with chalk or powdered gypsum or applied to paper. Litmus is used as a chemical indicator for acids and alkalis because its natural purple colour is change to red by acids and to blue by alkalis.

8

Essence of Botany

Bud

In botany, a bud is an undeveloped or embryonic shoot and normally occurs in the axil of a leaf or at the tip of the stem. Once formed, a bud may remain for some time in a dormant condition, or it may form a shoot immediately. Buds may be specialized to develop flowers or short shoots, or may have the potential for general shoot development. The term bud is also used in zoology, where it refers to an outgrowth from the body which can develop into a new individual.

***Figure:** Flower buds have not yet bloomed into a full-size flower*

Overview

The buds of many woody plants, especially in temperate or cold climates, are protected by a covering of modified leaves called *scales* which tightly enclose the more delicate parts of the bud. Many bud scales are covered by a gummy substance which serves as added protection. When the bud develops, the scales may enlarge somewhat but usually just drop off, leaving on the surface of the growing stem a series of horizontally-elongated scars.

By means of these scars one can determine the age of any young branch, since each year's growth ends in the formation of a bud, the formation of which produces an additional group of bud scale scars. Continued growth of the branch causes these scars to be obliterated after a few years so that the total age of older branches cannot be determined by this means.

Figure: *On the left is an opening influorescence bud, that will develop like the one to the right.*

In many plants scales are not formed over the bud, which is then called a naked bud. The minute underdeveloped leaves in such buds are often excessively hairy. Naked buds are found in some shrubs, like some species of the Sumac and Viburnums (*Viburnum alnifolium* and *V. lantana*) and in herbaceous plants.

In many of the latter, buds are even more reduced, often consisting of undifferentiated masses of cells in the axils of leaves. A terminal bud occurs on the end of a stem and lateral buds are found on the side. A head of cabbage is an exceptionally large terminal bud, while Brussels sprouts are large lateral buds.

Since buds are formed in the axils of leaves, their distribution on the stem is the same as that of leaves. There are alternate, opposite, and whorled buds, as well as the terminal bud at the tip of the stem. In many plants buds appear in unexpected places: these are known as adventitious buds.

Often it is possible to find a bud in a remarkable series of gradations of bud scales. In the buckeye, for example, one may see a complete gradation from the small brown outer scale through larger scales which on unfolding become somewhat green to the inner scales of the bud, which are remarkably leaf-like.

Such a series suggests that the scales of the bud are in truth leaves, modified to protect the more delicate parts of the plant during unfavourable periods.

Types of Buds

Buds are often useful in the identification of plants, especially for woody plants in winter when leaves have fallen. Buds may be classified and described according to different criteria: location, status, morphology, and function.

Botanists commonly use the following terms:

- for location:
  - o terminal, when located at the tip of a stem (apical is equivalent but rather reserved for the one at the top of the plant);
  - o axillary, when located in the axil of a leaf (lateral is the equivalent but some adventitious buds may be lateral too);
  - o adventitious, when occurring elsewhere, for example on trunk or on roots (some adventitious buds may be former axillary ones reduced and hidden under the bark, other adventitious buds are completely new formed ones).
- for status:
  - o accessory, for secondary buds formed besides a principal bud (axillary or terminal);
  - o resting, for buds that form at the end of a growth season, which will lie dormant until onset of the next growth season;
  - o dormant or latent, for buds whose growth has been delayed for a rather long time. The term is usable as a synonym of *resting*, but is rather employed for buds waiting undeveloped for years, for example epicormic buds;
  - o pseudoterminal, for an axillary bud taking over the function of a terminal bud (characteristic of species whose growth is sympodial: terminal bud dies and is replaced by the closer axillary bud, for examples beech, persimmon, *Platanus* have sympodial growth).
- for morphology:
  - o naked, when not covered by scales;

  - o scaly or covered, when scales (which are in fact transformed and reduced leaves) cover and protect the embryonic parts;
  - o hairy, when also protected by hairs (it may apply either to scaly or to naked buds).
- for function:
  - o vegetative, if only containing vegetative pieces: embryonic shoot with leaves (a leaf bud is the same);
  - o reproductive, if containing embryonic flower(s) (a flower bud is the same);
  - o mixed, if containing both embryonic leaves and flowers.

Within Zoology

The term bud (as in budding) is used by analogy within zoology as well, where it refers to an outgrowth from the body which develops into a new individual. It is a form of asexual reproduction limited to animals or plants of relatively simple structure. In this process a portion of the wall of the parent cell softens and pushes out. The protuberance thus formed enlarges rapidly while at this time the nucleus of the parent cell divides. One of the resulting nuclei passes into the bud, and then the bud is cut off from its parent cell and the process is repeated. Often the daughter cell will begin to bud before it becomes separated from the parent, so that whole colonies of adhering cells may be formed. Eventually cross walls cut off the bud from the original cell.

Chlorophyll

Chlorophyll (also chlorophyl) is a green pigment found in almost all plants, algae, and cyanobacteria. Chlorophyll is an extremely important biomolecule, critical in photosynthesis, which allows plants to absorb energy from light. Chlorophyll absorbs light most strongly in the blue portion of the electromagnetic spectrum, followed by the red portion. However, it is a poor absorber of green and near-green portions of the spectrum, hence the green colour of chlorophyll-containing tissues. Chlorophyll was first isolated by Joseph Bienaimé Caventou and Pierre Joseph Pelletier in 1817.

Chlorophyll and Photosynthesis

Chlorophyll is vital for photosynthesis, which allows plants to absorb energy from light. Chlorophyll molecules are specifically arranged in and around photosystems that are embedded in the thylakoid membranes of chloroplasts. In these complexes, chlorophyll serves two primary functions. The function of the vast majority of chlorophyll (up to several hundred molecules per photosystem) is to absorb light and transfer that light energy by resonance energy transfer to a specific chlorophyll pair in the reaction centre of the photosystems.

The two currently accepted photosystem units are Photosystem II and Photosystem I, which have their own distinct reaction centre chlorophylls, named P680 and P700, respectively. These pigments are named after the wavelength (in nanometers) of their red-peak absorption maximum. The identity, function and spectral properties of the types of chlorophyll in each photosystem are distinct and determined by each other and the protein structure surrounding them. Once extracted from the protein into a solvent (such as acetone or methanol), these chlorophyll pigments can be separated in a simple paper chromatography experiment and, based on the number of polar groups between chlorophyll a and chlorophyll b, will chemically separate out on the paper.

The function of the reaction centre chlorophyll is to use the energy absorbed by and transferred to it from the other chlorophyll pigments in the photosystems to undergo a charge separation, a specific redox reaction in which the chlorophyll donates an electron into a series of molecular intermediates called an electron transport chain.

The charged reaction centre chlorophyll ($P680^+$) is then reduced back to its ground state by accepting an electron. In Photosystem II, the electron that reduces $P680^+$ ultimately comes from the oxidation of water into O_2 and H^+ through several intermediates.

This reaction is how photosynthetic organisms such as plants produce O_2 gas, and is the source for practically all the O_2 in Earth's atmosphere. Photosystem I typically works in series with Photosystem II; thus the $P700^+$ of Photosystem I is usually reduced, via many intermediates in the thylakoid membrane, by electrons ultimately from Photosystem II. Electron transfer reactions in the thylakoid membranes are complex, however, and the source of electrons used to reduce $P700^+$ can vary.

The electron flow produced by the reaction centre chlorophyll pigments is used to shuttle H^+ ions across the thylakoid membrane, setting up a chemiosmotic potential used mainly to produce ATP chemical energy; and those electrons ultimately reduce $NADP^+$ to NADPH, a universal reductant used to reduce CO_2 into sugars as well as for other biosynthetic reductions.

Reaction centre chlorophyll–protein complexes are capable of directly absorbing light and performing charge separation events without other chlorophyll pigments, but the absorption cross section (the likelihood of absorbing a photon under a given light intensity) is small.

Thus, the remaining chlorophylls in the photosystem and antenna pigment protein complexes associated with the photosystems all cooperatively absorb and funnel light energy to the reaction centre. Besides chlorophyll *a*, there are other pigments, called accessory pigments, which occur in these pigment–protein antenna complexes. A green sea slug, *Elysia chlorotica*, has been found to use the chlorophyll it has eaten to perform photosynthesis for itself. This process is known as kleptoplasty, and no other animal has been found to have this ability.

Why Green and not Black?

It still is unclear exactly why plants have mostly evolved to be green. Green plants reflect mostly green and near-green light to viewers rather than absorbing it. Other parts of the system of photosynthesis still allow green plants to use the green light spectrum (e.g., through a light-trapping leaf structure, carotenoids, etc.). Green plants do not use a large part of the visible spectrum as efficiently as possible.

A black plant can absorb more radiation, and this could be very useful, if extra heat produced is effectively disposed of (e.g., some plants must close their openings, called stomata, on hot days to avoid losing too much water, which leaves only conduction, convection, and radiative heat-loss as solutions). The question becomes why the only light-absorbing molecule used for power in plants is green and not simply black.

The biologist John Berman has offered the opinion that evolution is not an engineering process, and so it is often subject to various limitations that an engineer or other designer is not. Even *if* black leaves were better, evolution's limitations can prevent species from climbing to the absolute highest peak on the fitness landscape.

Berman wrote that achieving pigments that work better than chlorophyll could be very difficult. In fact, all higher plants (embryophytes) are thought to have evolved from a common ancestor that is a sort of green algae – with the idea being that chlorophyll has evolved only once.

Shil DasSarma, a microbial geneticist at the University of Maryland, has pointed out that species of archaea do use another light-absorbing molecule, retinal, to extract power from the green spectrum. He described the view of some scientists that such green-light-absorbing archae once dominated the earth environment.

This could have left open a "niche" for green organisms that would absorb the other wavelengths of sunlight. This is just a possibility, and Berman wrote that scientists are still not convinced of any one explanation.

Chemical Structure

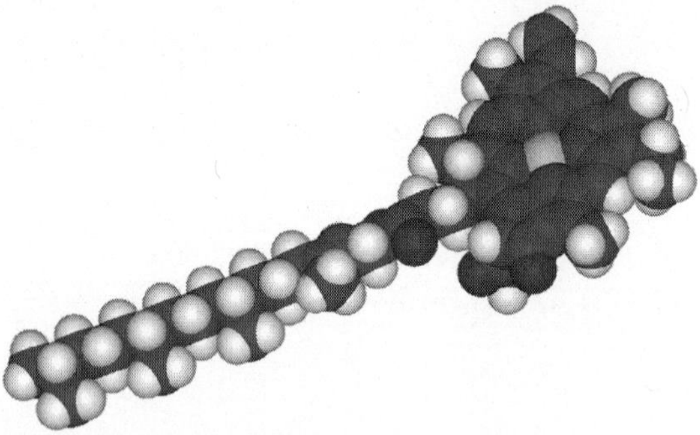

Figure: *Space-filling model of the chlorophyll a molecule*

Chlorophyll is a chlorin pigment, which is structurally similar to and produced through the same metabolic pathway as other porphyrin pigments such as heme. At the centre of the chlorin ring is a magnesium ion.

At the time of its discovery in the early 1900s, this was the first time that this element had been detected in living tissue. For the structures depicted in this article, some of the ligands attached to the Mg^{2+} centre are omitted for clarity.

The chlorin ring can have several different side chains, usually including a long phytol chain. There are a few different forms that occur naturally, but the most widely distributed form in terrestrial plants is chlorophyll *a*. After initial work done by German chemist Richard Willstätter spanning from 1905 to 1915, the general structure of chlorophyll *a* was elucidated by Hans Fischer in 1940.

By 1960, when most of the stereochemistry of chlorophyll *a* was known, Robert Burns Woodward published a total synthesis of the molecule. In 1967, the last remaining stereochemical elucidation was completed by Ian Fleming, and in 1990 Woodward and co-authors published an updated synthesis.

Chlorophyll was announced to be present in cyanobacteria and other oxygenic microorganisms that form stromatolites in 2010; a molecular formula of $C_{55}H_{70}O_6N_4Mg$ and a structure of (2-formyl)-chlorophyll *a* were deduced based on NMR, optical and mass spectra.

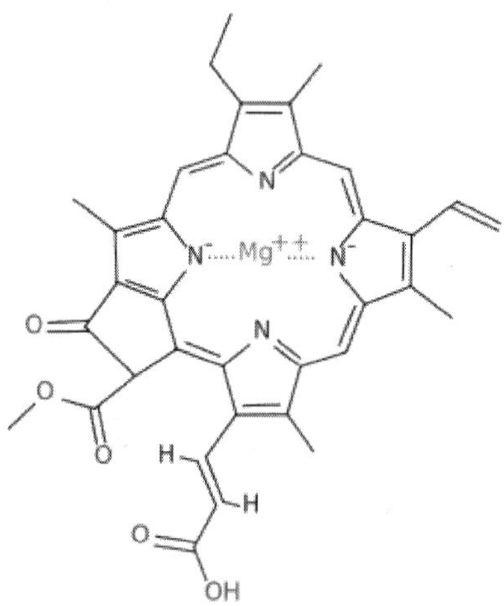

Figure: *Structure of Chlorophyll c1*

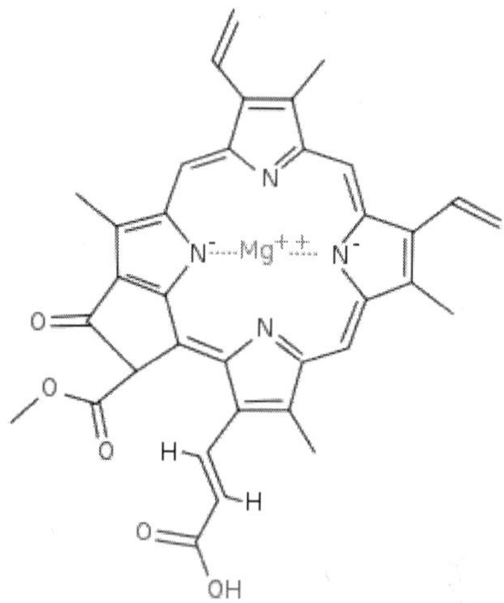

Figure: Structure of Chlorophyll c2

When leaves degreen in the process of plant senescence, chlorophyll is converted to a group of colourless tetrapyrroles known as nonfluorescent chlorophyll catabolites (NCC's) with the general structure:

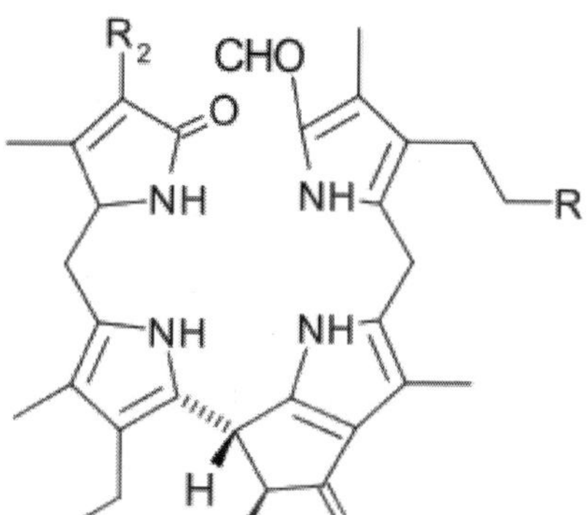

Figure: *These compounds have also been identified in several ripening fruits.*

Spectrophotometry

Measurement of the absorption of light is complicated by the solvent used to extract it from plant material, which affects the values obtained,

- In diethyl ether, chlorophyll *a* has approximate absorbance maxima of 430 nm and 662 nm, while chlorophyll *b* has approximate maxima of 453 nm and 642 nm.
- The absorption peaks of chlorophyll *a* are at 665 nm and 465 nm. Chlorophyll *a* fluoresces at 673 nm (maximum) and 726 nm. The peak molar absorption coefficient of chlorophyll *a* exceeds 10^5 M^{-1} cm^{-1}, which is among the highest for small-molecule organic compounds.
- In 90% acetone-water, the peak absorption wavelengths of chlorophyll *a* are 430 nm and 664 nm; peaks for chlorophyll *b* are 460 nm and 647 nm; peaks for chlorophyll *c1* are 442 nm and 630 nm; peaks for chlorophyll *c2* are 444 nm and 630 nm; peaks for chlorophyll *d* are 401 nm, 455 nm and 696 nm.

By measuring the absorption of light in the red and far red regions it is possible to estimate the concentration of chlorophyll within a leaf.

While absorption instruments have been proven over time to work well as a non-destructive test on many types of samples, they do have limitations. Absorption technique limitations:

- The sample to be measured must completely cover the instrument's aperture, with no holes.
- Samples must be thin enough to allow transmission of the measuring wavelengths of light.
- Surfaces must be relatively flat and uniform.
- Variable fluorescence (the Kautsky induction effect) caused by the red wavelength of absorption instruments, limits the repeatability of measurements at the same location.
- Selection of the measuring area, on smaller leaves, can cause measurement variations due to internal leaf structure variation. Veins and midribs can cause significant variation on samples.
- Wavelengths used in absorption methods, typically limit linear correlation with chemical chlorophyll content methods to concentration levels below 300 mg/m^2.

As a result, absorption instruments do not work with conifer needles, turf grasses, Arabidopsis leaves, moss, most CAM plants such as prickly pear cactus, and Agave, fruit, stems, petioles, lichens, and algae on rocks. Furthermore, it is difficult to get reliable readings on very small leaf plants such as immature rice and wheat.

A superior method is to measure the chlorophyll content using chlorophyll fluorescence. Gitelson (1999) states, "The ratio between chlorophyll fluorescence, at 735 nm and the wavelength range 700nm to 710 nm, F735/F700 was found to be linearly proportional to the chlorophyll content (with determination coefficient, r2, more than 0.95) and thus this ratio can be used as a precise indicator of chlorophyll content in plant leaves."

Non-destructive, hand-held meters use this effect to measure the chlorophyll content of samples where traditional absorbance techniques cannot be used.

This new method has many advantages over older, absorption techniques. These include:-

1. Samples do not need to fill the measuring aperture, so small samples can be measured.
2. Samples do not need to allow transmission of the measuring beam, so thick samples can be measured.
3. Surfaces do not need to be flat and uniform, so difficult plant morphologies can be measured.
4. No Kautsky induction effect, so measurements can be repeated rapidly at the same sight.
5. Small sample area, so naturally occurring structures like mid ribs and veins can be avoided.
6. Higher measuring range (675 mg/m^2, as opposed to 300 mg/m^2 with absorption techniques), so many more samples can be measured.

By measuring chlorophyll fluorescence, plant ecophysiology can be investigated. Chlorophyll fluorometers are used by plant researchers to assess plant stress.

Biosynthesis

In plants, chlorophyll may be synthesized from succinyl-CoA and glycine, although the immediate precursor to chlorophyll *a* and *b* is protochlorophyllide. In Angiosperm plants, the last step, conversion of protochlorophyllide to chlorophyll, is light-dependent and such plants are pale (etiolated) if grown in the darkness. Non-vascular plants and green algae have an additional light-independent enzyme and grow green in the darkness instead.

Chlorophyll itself is bound to proteins and can transfer the absorbed energy in the required direction. Protochlorophyllide occurs mostly in the free form and, under light conditions, acts as a photosensitizer, forming highly toxic free radicals. Hence, plants need an efficient mechanism of regulating the amount of chlorophyll precursor. In angiosperms, this is done at the step of aminolevulinic acid (ALA), one of the intermediate compounds in the biosynthesis pathway. Plants that are fed by ALA accumulate high and toxic levels of protochlorophyllide; so do the mutants with the damaged regulatory system.

Chlorosis is a condition in which leaves produce insufficient chlorophyll, turning them yellow. Chlorosis can be caused by a nutrient deficiency of iron—called iron chlorosis—or by a shortage of magnesium or nitrogen. Soil pH sometimes plays a role in nutrient-caused chlorosis; many plants are adapted to grow in soils with specific pH levels and their ability to absorb nutrients from the soil can be dependent on this. Chlorosis can also be caused by pathogens including viruses, bacteria and fungal infections, or sap-sucking insects.

Complementary Light Absorbance of Anthocyanins with Chlorophylls

Anthocyanins are other plant pigments. The absorbance pattern responsible for the red colour of anthocyanins may be complementary to that of green chlorophyll in photosynthetically active tissues such as young *Quercus coccifera* leaves. It may protect the leaves from attacks by plant eaters that may be attracted by green colour.

Culinary Use

Chlorophyll is registered as a food additive (colourant), and its E number is E140. Chefs use chlorophyll to colour a variety of foods and beverages green, such as pasta and absinthe. Chlorophyll is not soluble in water, and it is first mixed with a small quantity of vegetable oil to obtain the desired solution. Extracted liquid chlorophyll was considered to be unstable and always denatured until 1997, when Frank S. & Lisa Sagliano used freeze-drying of liquid chlorophyll at the University of Florida and stabilized it as a powder, preserving it for future use.

Chloroplast

Chloroplasts are organelles found in plant cells and other eukaryotic organisms that conduct photosynthesis. Chloroplasts capture light energy, store it in the energy storage molecules ATP and NADPH and use it in the process called photosynthesis to make organic molecules and free oxygen from carbon dioxide and water.

Chloroplasts are green because they contain the chlorophyll pigment. The word *chloroplast* is derived from the Greek words *chloros*, which means green, and *plastis*, which means "the one who forms". Chloroplasts are members of a class of organelles known as plastids.

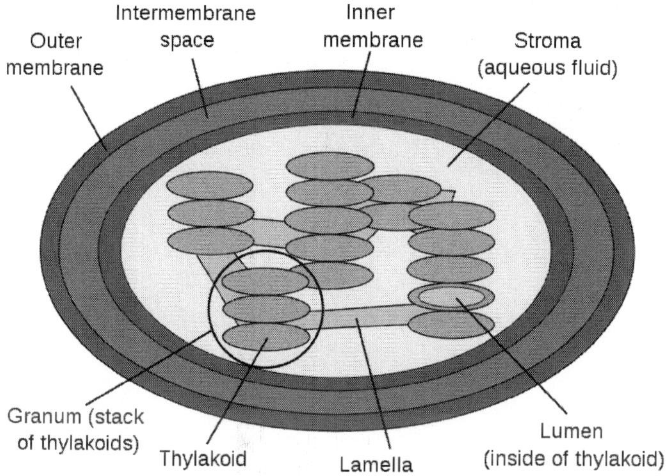

Figure: *Simplified Structure of a Chloroplast*

Evolutionary Origin

Chloroplasts are one of the many different types of organelles in the plant cell. They are considered to have originated from cyanobacteria through endosymbiosis. This was first suggested by Mereschkowsky in 1905 after an observation by Schimper in 1883 that chloroplasts closely resemble cyanobacteria. All chloroplasts are thought to derive directly or indirectly from a single endosymbiotic event (in the Archaeplastida), except for *Paulinella chromatophora*, which has recently acquired a photosynthetic cyanobacterial endosymbiont which is not closely related to chloroplasts of other eukaryotes.

In that they derive from an endosymbiotic event, chloroplasts are similar to mitochondria, but chloroplasts are found only in plants and protista. The chloroplast is surrounded by a double-layered composite membrane with an intermembrane space; further, it has reticulations, or many infoldings, filling the inner spaces. The chloroplast has its own DNA, which codes for redox proteins involved in electron transport in photosynthesis; this is termed the plastome.

In green plants, chloroplasts are surrounded by two lipid-bilayer membranes that are thought to correspond to the outer and inner membranes of the ancestral cyanobacterium. Chloroplasts have their own genome, which is considerably reduced compared to that of free-living cyanobacteria, but the parts that are still present show clear similarities with the cyanobacterial genome. Plastids may contain 60-100 genes whereas cyanobacteria often contain more than 1500 genes. Many of the missing genes are encoded in the nuclear genome of the host.

In some algae (such as the heterokonts and other protists such as Euglenozoa and Cercozoa), chloroplasts seem to have evolved through a secondary event of endosymbiosis, in which a eukaryotic cell engulfed a second eukaryotic cell containing chloroplasts, forming chloroplasts with three or four membrane layers. In some cases, such secondary endosymbionts may have themselves been engulfed by still other eukaryotes, thus forming tertiary endosymbionts. In the alga *Chlorella*, there is only one chloroplast, which is bell-shaped.

In some groups of mixotrophic protists such as the dinoflagellates, chloroplasts are separated from a captured alga or diatom and used temporarily. These klepto chloroplasts may only have a lifetime of a few days and are then replaced.

Structure

Chloroplasts are observable as flat discs usually 2 to 10 micrometers in diameter and 1 micrometer thick. In land plants, they are, in general, 5 ìm in diameter and 2.3 μm thick. The chloroplast is contained by an envelope that consists of an inner and an outer phospholipid membrane. Between these two layers is the intermembrane space. A typical parenchyma cell contains about 10 to 100 chloroplasts. The material within the chloroplast is called the stroma, corresponding to the cytosol of the original bacterium, and contains one or more molecules of small circular DNA. It also contains ribosomes; however most of its proteins are encoded by genes contained in the host cell nucleus, with the protein products transported to the chloroplast.

Within the stroma are stacks of thylakoids, the sub-organelles, which are the site of photosynthesis. The thylakoids are arranged in stacks called grana (singular: granum). A thylakoid has a flattened disk shape. Inside it is an empty area called the thylakoid space or lumen. Photosynthesis takes place on the thylakoid membrane; as in mitochondrial oxidative phosphorylation, it involves the coupling of cross-membrane fluxes with biosynthesis via the dissipation of a proton electrochemical gradient.

In the electron microscope, thylakoid membranes appear as alternating light-and-dark bands, each 0.01 ìm thick. Embedded in the thylakoid membrane are antenna complexes, each of which consists of the light-absorbing pigments, including chlorophyll and carotenoids, as well as proteins that bind the pigments.

This complex both increases the surface area for light capture, and allows capture of photons with a wider range of wavelengths. The energy of the incident photons is absorbed by the pigments and funneled to the reaction centre of this complex through resonance energy transfer. Two chlorophyll molecules are then ionised, producing an excited electron, which then passes onto the photochemical reaction centre.

Recent studies have shown that chloroplasts can be interconnected by tubular bridges called stromules, formed as extensions of their outer membranes. Chloroplasts appear to be able to exchange proteins via stromules, and thus function as a network.

Transplastomic Plants

Recently, chloroplasts have caught attention by developers of genetically modified plants. In most flowering plants, chloroplasts are not inherited from the male parent, although in plants such as pines, chloroplasts are inherited from males. Where chloroplasts are inherited only from the female, transgenes in these plastids cannot be disseminated by pollen.

This makes plastid transformation a valuable tool for the creation and cultivation of genetically modified plants that are biologically contained, thus posing significantly lower environmental risks.

This biological containment strategy is therefore suitable for establishing the coexistence of conventional and organic agriculture.

While the reliability of this mechanism has not yet been studied for all relevant crop species, recent results in tobacco plants are promising, showing a failed containment rate of transplastomic plants at 3 in 1,000,000.

Flora

"Flora" comes from the Latin name of Flora, the goddess of plants, flowers, and fertility in Roman mythology.

Flora is the plant life occurring in a particular region or time, generally the naturally occurring or indigenous—native plant life. The corresponding term for animal life is fauna. *Flora, fauna* and other forms of life such as fungi are collectively referred to as biota.

Flower

A flower, sometimes known as a bloom or blossom, is the reproductive structure found in flowering plants (plants of the division Magnoliophyta, also called angiosperms). The biological function of a flower is to effect reproduction, usually by providing a mechanism for the union of sperm with eggs. Flowers may facilitate outcrossing (fusion of sperm and eggs from different individuals in a population) or allow selfing (fusion of sperm and egg from the same flower).

Some flowers produce diaspores without fertilization (parthenocarpy). Flowers contain sporangia and are the site where gametophytes develop. Flowers give rise to fruit and seeds. Many flowers have evolved to be attractive to animals, so as to cause them to be vectors for the transfer of pollen. In addition to facilitating the reproduction of flowering plants, flowers have long been admired and used by humans to beautify their environment, and also as objects of romance, ritual, religion, medicine and as a source of food.

Morphology

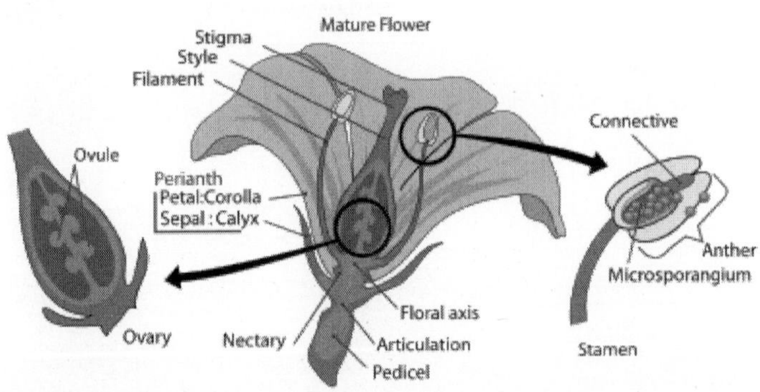

***Figure:** Diagram showing the main parts of a mature flower*

A stereotypical flower consists of four kinds of structures attached to the tip of a short stalk. Each of these kinds of parts is arranged in a whorl on the receptacle. The four main whorls (starting from the base of the flower or lowest node and working upwards) are as follows:

- *Calyx*: the outermost whorl consisting of units called *sepals*; these are typically green and enclose the rest of the flower in the bud stage, however, they can be absent or prominent and petal-like in some species.
- *Corolla*: the next whorl towards the apex, composed of units called *petals*, which are typically thin, soft and coloured to attract animals that help the process of pollination.

- *Androecium*: the next whorl (sometimes multiplied into several whorls), consisting of units called stamens. Stamens consist of two parts: a stalk called a filament, topped by an anther where pollen is produced by meiosis and eventually dispersed.
- *Gynoecium*: the innermost whorl of a flower, consisting of one or more units called carpels. The carpel or multiple fused carpels form a hollow structure called an ovary, which produces ovules internally. Ovules are megasporangia and they in turn produce megaspores by meiosis which develop into female gametophytes. These give rise to egg cells.

  The gynoecium of a flower is also described using an alternative terminology wherein the structure one sees in the innermost whorl (consisting of an ovary, style and stigma) is called a pistil. A pistil may consist of a single carpel or a number of carpels fused together. The sticky tip of the pistil, the stigma, is the receptor of pollen. The supportive stalk, the style, becomes the pathway for pollen tubes to grow from pollen grains adhering to the stigma.

Although the arrangement described above is considered "typical", plant species show a wide variation in floral structure. These modifications have significance in the evolution of flowering plants and are used extensively by botanists to establish relationships among plant species.

The four main parts of a flower are generally defined by their positions on the receptacle and not by their function. Many flowers lack some parts or parts may be modified into other functions and/or look like what is typically another part.

In some families, like Ranunculaceae, the petals are greatly reduced and in many species the sepals are colourful and petal-like. Other flowers have modified stamens that are petal-like, the double flowers of Peonies and Roses are mostly petaloid stamens. Flowers show great variation and plant scientists describe this variation in a systematic way to identify and distinguish species.

Specific terminology is used to describe flowers and their parts. Many flower parts are fused together; fused parts originating from the same whorl are connate, while fused parts originating from different whorls are adnate, parts that are not fused are free. When petals are fused into a tube or ring that falls away as a single unit, they are sympetalous (also called gamopetalous.)

Connate petals may have distinctive regions: the cylindrical base is the tube, the expanding region is the throat and the flaring outer region is the limb. A sympetalous flower, with bilateral symmetry with an upper and lower lip, is bilabiate. Flowers with connate petals or sepals may have various shaped corolla or calyx including: campanulate, funnelform, tubular, urceolate, salverform or rotate.

Many flowers have a symmetry. When the perianth is bisected through the central axis from any point, symmetrical halves are produced, forming a radial symmetry.

These flowers are also known to be actinomorphic or regular, e.g. rose or trillium. When flowers are bisected and produce only one line that produces symmetrical halves the flower is said to be irregular or zygomorphic, e.g. snapdragon or most orchids.

Flowers may be directly attached to the plant at their base (sessile—the supporting stalk or stem is highly reduced or absent). The stem or stalk subtending a flower is called a peduncle.

If a peduncle supports more than one flower, the stems connecting each flower to the main axis are called pedicels. The apex of a flowering stem forms a terminal swelling which is called the *torus* or receptacle.

Floral Formula

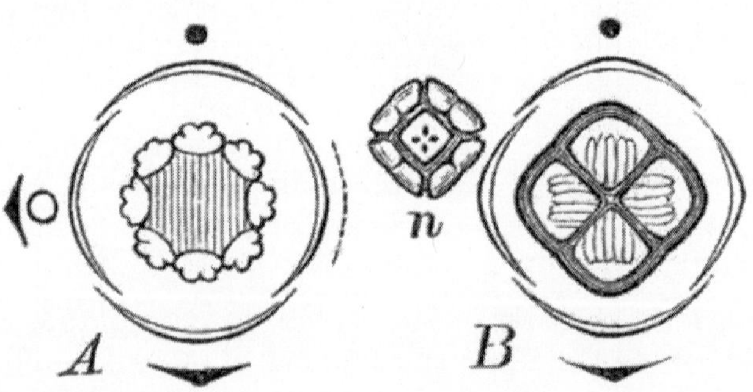

Figure: Floral diagram for the genus Nepenthes

A *floral formula* is a way to represent the structure of a flower using specific letters, numbers, and symbols. Typically, a general formula will be used to represent the flower structure of a plant family rather than a particular species. The following representations are used:

Ca = calyx (sepal whorl; e. g. Ca^5 = 5 sepals)

Co = corolla (petal whorl; e. g., $Co^{3(x)}$ = petals some multiple of three)

Z = add if *zygomorphic* (e. g., CoZ^6 = zygomorphic with 6 petals)

A = *androecium* (whorl of stamens; e. g., A" = many stamens)

G = *gynoecium* (carpel or carpels; e. g., G^1 = monocarpous)

x: to represent a "variable number"

∞: to represent "many"

A floral formula would appear something like this:

$Ca^5Co^5A^{10-\infty}G^1$

Several additional symbols are sometimes used.

Influorescence

In those species that have more than one flower on an axis, the collective cluster of flowers is termed an *influorescence.* Some influorescences are composed of many small flowers arranged in a formation that resembles a single flower. The common example of this is most members of the very large composite (Asteraceae) group. A single daisy or sunflower, for example, is not a flower but a flower *head*—an influorescence composed of numerous flowers (or florets).

An influorescence may include specialized stems and modified leaves known as bracts.

Development

A flower is a modified stem tip with compressed internodes, bearing structures that are highly modified leaves. In essence, a flower develops on a modified shoot or *axis* from a determinate apical meristem (*determinate* meaning the axis grows to a set size).

Flowering Transition

The transition to flowering is one of the major phase changes that a plant makes during its life cycle. The transition must take place at a time that is favourable for fertilization and the formation of seeds, hence ensuring maximal reproductive success. To meet these needs a plant is able to interpret important endogenous and environmental cues such as changes in levels of plant hormones and seasonable temperature and photoperiod changes. Many perennial and most biennial plants require vernalization to flower. The molecular interpretation of these signals is through the transmission of a complex signal known as florigen, which involves a variety of genes, including CONSTANS, FLOWERING LOCUS C and FLOWERING LOCUS T. Florigen is produced in the leaves in reproductively favourable conditions and acts in buds and growing tips to induce a number of different physiological and morphological changes.

The first step is the transformation of the vegetative stem primordia into floral primordia. This occurs as biochemical changes take place to change cellular differentiation of leaf, bud and stem tissues into tissue that will grow into the reproductive organs. Growth of the central part of the stem tip stops or flattens out and the sides develop protuberances in a whorled or spiral fashion around the outside of the stem end.

These protuberances develop into the sepals, petals, stamens, and carpels. Once this process begins, in most plants, it cannot be reversed and the stems develop flowers, even if the initial start of the flower formation event was dependent of some environmental cue. Once the process begins, even if that cue is removed the stem will continue to develop a flower.

Organ Development

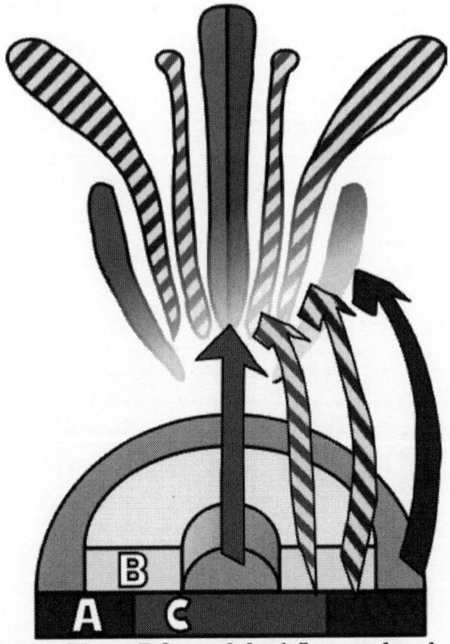

Figure: *The ABC model of flower development*

The molecular control of floral organ identity determination is fairly well understood. In a simple model, three gene activities interact in a combinatorial manner to determine the developmental identities of the organ primordia within the floral meristem.

These gene functions are called A, B and C-gene functions. In the first floral whorl only A-genes are expressed, leading to the formation of sepals. In the second whorl both A- and B-genes are expressed, leading to the formation of petals. In the third whorl, B and C genes interact to form stamens and in the centre of the flower C-genes alone give rise to carpels.

The model is based upon studies of homeotic mutants in *Arabidopsis thaliana* and snapdragon, *Antirrhinum majus*. For example, when there is a loss of B-gene function, mutant flowers are produced with sepals in the first whorl as usual, but also in the second whorl instead of the normal petal formation. In the third whorl the lack of B function but presence of C-function mimics the fourth whorl, leading to the formation of carpels also in the third whorl. See also The ABC Model of Flower Development. Most genes central in this model belong to the MADS-box genes and are transcription factors that regulate the expression of the genes specific for each floral organ.

Floral Function

The principal purpose of a flower is the reproduction of the individual and the species. All flowering plants are *heterosporous*, producing two types of spores. Microspores are produced by meiosis inside anthers while megaspores are produced inside ovules, inside an ovary.

In fact, anthers typically consist of four microsporangia and an ovule is an integumented megasporangium. Both types of spores develop into gametophytes inside sporangia. As with all heterosporous plants, the gametophytes also develop inside the spores (are endosporic).

In the majority of species, individual flowers have both functional carpels and stamens. These flowers are described by botanists as being *perfect* or *bisexual*. Some flowers lack one or the other reproductive organ and called *imperfect* or *unisexual* If unisex flowers are found on the same individual plant but in different locations, the species is said to be *monoecious*. If each type of unisex flower is found only on separate individuals, the plant is *dioecious*.

Flower Specialization and Pollination

Flowering plants usually face selective pressure to optimize the transfer of their pollen, and this is typically reflected in the morphology of the flowers and the behaviour of the plants. Pollen may be transferred between plants via a number of 'vectors'. Some plants make use of abiotic vectors — namely wind (anemophily) or, much less commonly, water (hydrophily). Others use biotic vectors including insects (entomophily), birds (ornithophily), bats (chiropterophily) or other animals. Some plants make use of multiple vectors, but many are highly specialised.

Cleistogamous flowers are self pollinated, after which they may or may not open. Many Viola and some Salvia species are known to have these types of flowers.

The flowers of plants that make use of biotic pollen vectors commonly have glands called nectaries that act as an incentive for animals to visit the flower. Some flowers have patterns, called nectar guides, that show pollinators where to look for nectar. Flowers also attract pollinators by scent and colour. Still other flowers use mimicry to attract pollinators. Some species of orchids, for example, produce flowers resembling female bees in colour, shape, and scent. Flowers are also specialized in shape and have an arrangement of the stamens that ensures that pollen grains are transferred to the bodies of the pollinator when it lands in search of its attractant (such as nectar, pollen, or a mate). In pursuing this attractant from many flowers of the same species, the pollinator transfers pollen to the stigmas—arranged with equally pointed precision—of all of the flowers it visits.

Anemophilous flowers use the wind to move pollen from one flower to the next. Examples include grasses, birch trees, ragweed and maples. They have no need to attract pollinators and therefore tend not to be "showy" flowers. Male and female reproductive organs are generally found in separate flowers, the male flowers having a number of long filaments terminating in exposed stamens, and the female flowers having long, feather-like stigmas.

Whereas the pollen of animal-pollinated flowers tends to be large-grained, sticky, and rich in protein (another "reward" for pollinators), anemophilous flower pollen is usually small-grained, very light, and of little nutritional value to animals.

Pollination

The primary purpose of a flower is reproduction. Since the flowers are the reproductive organs of plant, they mediate the joining of the sperm, contained within pollen, to the ovules — contained in the ovary. Pollination is the movement of pollen from the anthers to the stigma. The joining of the sperm to the ovules is called fertilization. Normally pollen is moved from one plant to another, but many plants are able to self pollinate. The fertilized ovules produce seeds that are the next generation. Sexual reproduction produces genetically unique offspring, allowing for adaptation. Flowers have specific designs which encourages the transfer of pollen from one plant to another of the same species. Many plants are dependent upon external factors for pollination, including: wind and animals, and especially insects. Even large animals such as birds, bats, and pygmy possums can be employed. The period of time during which this process can take place (the flower is fully expanded and functional) is called *anthesis*.

Attraction Methods

Plants cannot move from one location to another, thus many flowers have evolved to attract animals to transfer pollen between individuals in dispersed populations.

Figure: *A Bee orchid has evolved over many generations to better mimic a female bee to attract male bees as pollinators.*

Flowers that are insect-pollinated are called *entomophilous*; literally "insect-loving" in Greek. They can be highly modified along with the pollinating insects by co-evolution. Flowers commonly have glands called *nectaries* on various parts that attract animals looking for nutritious nectar. Birds and bees have colour vision, enabling them to seek out "colourful" flowers. Some flowers have patterns, called nectar guides, that show pollinators where to look for nectar; they may be visible only under ultraviolet light, which is visible to bees and some other insects. Flowers also attract pollinators by scent and some of those scents are pleasant to our sense of smell. Not all flower scents are appealing to humans; a number of flowers are pollinated by insects that are attracted to rotten flesh and have flowers that smell like dead animals, often called Carrion flowers, including *Rafflesia*, the titan arum, and the North American pawpaw (*Asimina triloba*). Flowers pollinated by night visitors, including bats and moths, are likely to concentrate on scent to attract pollinators and most such flowers are white.

Still other flowers use mimicry to attract pollinators. Some species of orchids, for example, produce flowers resembling female bees in colour, shape, and scent. Male bees move from one such flower to another in search of a mate.

Pollination Mechanism

The pollination mechanism employed by a plant depends on what method of pollination is utilized.

Most flowers can be divided between two broad groups of pollination methods:

Entomophilous: flowers attract and use insects, bats, birds or other animals to transfer pollen from one flower to the next. Often they are specialized in shape and have an arrangement of the stamens that ensures that pollen grains are transferred to the bodies of the pollinator when it lands in search of its attractant (such as nectar, pollen, or a mate).

In pursuing this attractant from many flowers of the same species, the pollinator transfers pollen to the stigmas—arranged with equally pointed precision—of all of the flowers it visits. Many flowers rely on simple proximity between flower parts to ensure pollination. Others, such as the *Sarracenia* or lady-slipper orchids, have elaborate designs to ensure pollination while preventing self-pollination.

Anemophilous: flowers use the wind to move pollen from one flower to the next, examples include the grasses, Birch trees, Ragweed and Maples. They have no need to attract pollinators and therefore tend not to be "showy" flowers. Whereas the pollen of entomophilous flowers tends to be large-grained, sticky, and rich in protein (another "reward" for pollinators), anemophilous flower pollen is usually small-grained, very light, and of little nutritional value to insects, though it may still be gathered in times of dearth. Honeybees and bumblebees actively gather anemophilous corn (maize) pollen, though it is of little value to them.

Some flowers are self pollinated and use flowers that never open or are self pollinated before the flowers open, these flowers are called cleistogamous. Many Viola species and some Salvia have these types of flowers.

Flower-pollinator Relationships

Many flowers have close relationships with one or a few specific pollinating organisms. Many flowers, for example, attract only one specific species of insect, and therefore rely on that insect for successful reproduction. This close relationship is often given as an example of coevolution, as the flower and pollinator are thought to have developed together over a long period of time to match each other's needs.

This close relationship compounds the negative effects of extinction. The extinction of either member in such a relationship would mean almost certain extinction of the other member as well. Some endangered plant species are so because of shrinking pollinator populations.

Fertilization and Dispersal

Some flowers with both stamens and a pistil are capable of self-fertilization, which does increase the chance of producing seeds but limits genetic variation. The extreme case of self-fertilization occurs in flowers that always self-fertilize, such as many dandelions. Conversely, many species of plants have ways of preventing self-fertilization. Unisexual male and female flowers on the same plant may not appear or mature at the same time, or pollen from the same plant may be incapable of fertilizing its ovules. The latter flower types, which have chemical barriers to their own pollen, are referred to as self-sterile or self-incompatible.

Evolution

Figure: *Amborella trichopoda, the sister group to the rest of the flowering plants*

While land plants have existed for about 425 million years, the first ones reproduced by a simple adaptation of their aquatic counterparts: spores. In the sea, plants—and some animals—can simply scatter out genetic clones of themselves to float away and grow elsewhere. This is how early plants reproduced. But plants soon evolved methods of protecting these copies to deal with drying out and other abuse which is even more likely on land than in the sea. The protection became the seed, though it had not yet evolved the flower. Early seed-bearing plants include the ginkgo and conifers. The earliest fossil of a flowering plant, *Archaefructus liaoningensis*, is dated about 125 million years old. Several groups of extinct gymnosperms, particularly seed ferns, have been proposed as the ancestors of flowering plants but there is no continuous fossil evidence showing exactly how flowers evolved.

The apparently sudden appearance of relatively modern flowers in the fossil record posed such a problem for the theory of evolution that it was called an "abominable mystery" by Charles Darwin. Recently discovered angiosperm fossils such as *Archaefructus*, along with further discoveries of fossil gymnosperms, suggest how angiosperm characteristics may have been acquired in a series of steps.

Recent DNA analysis (molecular systematics) shows that *Amborella trichopoda*, found on the Pacific island of New Caledonia, is the sister group to the rest of the flowering plants, and morphological studies suggest that it has features which may have been characteristic of the earliest flowering plants.

The general assumption is that the function of flowers, from the start, was to involve animals in the reproduction process. Pollen can be scattered without bright colours and obvious shapes, which would therefore be a liability, using the plant's resources, unless they provide some other benefit. One proposed reason for the sudden, fully developed appearance of flowers is that they evolved in an isolated setting like an island, or chain of islands, where the plants bearing them were able to develop a highly specialized relationship with some specific animal (a wasp, for example), the way many island species develop today.

This symbiotic relationship, with a hypothetical wasp bearing pollen from one plant to another much the way fig wasps do today, could have eventually resulted in both the plant(s) and their partners developing a high degree of specialization. Island genetics is believed to be a common source of speciation, especially when it comes to radical adaptations which seem to have required inferior transitional forms. Note that the wasp example is not incidental; bees, apparently evolved specifically for symbiotic plant relationships, are descended from wasps.

Likewise, most fruit used in plant reproduction comes from the enlargement of parts of the flower. This fruit is frequently a tool which depends upon animals wishing to eat it, and thus scattering the seeds it contains.

While many such symbiotic relationships remain too fragile to survive competition with mainland organisms, flowers proved to be an unusually effective means of production, spreading (whatever their actual origin) to become the dominant form of land plant life. While there is only hard proof of such flowers existing about 130 million years ago, there is some circumstantial evidence that they did exist up to 250 million years ago. A chemical used by plants to defend their flowers, oleanane, has been detected in fossil plants that old, including gigantopterids, which evolved at that time and bear many of the traits of modern, flowering plants, though they are not known to be flowering plants themselves, because only their stems and prickles have been found preserved in detail; one of the earliest examples of petrification.

The similarity in leaf and stem structure can be very important, because flowers are genetically just an adaptation of normal leaf and stem components on plants, a combination of genes normally responsible for forming new shoots. The most primitive flowers are thought to have had a variable number of flower parts, often separate from (but in contact with) each other.

The flowers would have tended to grow in a spiral pattern, to be bisexual (in plants, this means both male and female parts on the same flower), and to be dominated by the ovary (female part). As flowers grew more advanced, some variations developed parts fused together, with a much more specific number and design, and with either specific sexes per flower or plant, or at least "ovary inferior".

Flower evolution continues to the present day; modern flowers have been so profoundly influenced by humans that many of them cannot be pollinated in nature. Many modern, domesticated flowers used to be simple weeds, which only sprouted when the ground was disturbed. Some of them tended to grow with human crops, and the prettiest did not get plucked because of their beauty, developing a dependence upon and special adaptation to human affection.

Symbolism

Figure: *Lilies are often used to denote life or resurrection*

Many flowers have important symbolic meanings in Western culture. The practice of assigning meanings to flowers is known as floriography. Some of the more common examples include:

- Red roses are given as a symbol of love, beauty, and passion.
- Poppies are a symbol of consolation in time of death. In the United Kingdom, New Zealand, Australia and Canada, red poppies are worn to commemorate soldiers who have died in times of war.
- Irises/Lily are used in burials as a symbol referring to "resurrection/life". It is also associated with stars (sun) and its petals blooming/shining.

Flowers within art are also representative of the female genitalia, as seen in the works of artists such as Georgia O'Keeffe, Imogen Cunningham, Veronica Ruiz de Velasco, and Judy Chicago, and in fact in Asian and western classical art. Many cultures around the world have a marked tendency to associate flowers with femininity.

The great variety of delicate and beautiful flowers has inspired the works of numerous poets, especially from the 18th-19th century Romantic era. Famous examples include William Wordsworth's *I Wandered Lonely as a Cloud* and William Blake's *Ah! Sun-Flower*.

Because of their varied and colourful appearance, flowers have long been a favourite subject of visual artists as well. Some of the most celebrated paintings from well-known painters are of flowers, such as Van Gogh's sunflowers series or Monet's water lilies. Flowers are also dried, freeze dried and pressed in order to create permanent, three-dimensional pieces of flower art.

The Roman goddess of flowers, gardens, and the season of Spring is Flora. The Greek goddess of spring, flowers and nature is Chloris.

In Hindu mythology, flowers have a significant status. Vishnu, one of the three major gods in the Hindu system, is often depicted standing straight on a lotus flower. Apart from the association with Vishnu, the Hindu tradition also considers the lotus to have spiritual significance. For example, it figures in the Hindu stories of creation.

Usage

In modern times, people have sought ways to cultivate, buy, wear, or otherwise be around flowers and blooming plants, partly because of their agreeable appearance and smell. Around the world, people use flowers for a wide range of events and functions that, cumulatively, encompass one's lifetime:

- For new births or Christenings
- As a corsage or boutonniere to be worn at social functions or for holidays

- As tokens of love or esteem
- For wedding flowers for the bridal party, and decorations for the hall
- As brightening decorations within the home
- As a gift of remembrance for bon voyage parties, welcome home parties, and "thinking of you" gifts
- For funeral flowers and expressions of sympathy for the grieving
- For worshiping goddesses. in Hindu culture it is very common to bring flowers as a gift to temples.

People therefore grow flowers around their homes, dedicate entire parts of their living space to flower gardens, pick wildflowers, or buy flowers from florists who depend on an entire network of commercial growers and shippers to support their trade.

Flowers provide less food than other major plants parts (seeds, fruits, roots, stems and leaves) but they provide several important foods and spices. Flower vegetables include broccoli, cauliflower and artichoke. The most expensive spice, saffron, consists of dried stigmas of a crocus. Other flower spices are cloves and capers. Hops flowers are used to flavour beer. Marigold flowers are fed to chickens to give their egg yolks a golden yellow colour, which consumers find more desirable. Dandelion flowers are often made into wine. Bee Pollen, pollen collected from bees, is considered a health food by some people. Honey consists of bee-processed flower nectar and is often named for the type of flower, e.g. orange blossom honey, clover honey and tupelo honey.

Hundreds of fresh flowers are edible but few are widely marketed as food. They are often used to add colour and flavour to salads. Squash flowers are dipped in breadcrumbs and fried. Edible flowers include nasturtium, chrysanthemum, carnation, cattail, honeysuckle, chicory, cornflower, Canna, and sunflower. Some edible flowers are sometimes candied such as daisy and rose (you may also come across a candied pansy).

Flowers can also be made into herbal teas. Dried flowers such as chrysanthemum, rose, jasmine, camomile are infused into tea both for their fragrance and medical properties. Sometimes, they are also mixed with tea leaves for the added fragrance.

Flowers have been used since as far back as 50,000 years in funeral rituals. Many cultures do draw a connection between flowers and life and death, and because of their seasonal return flowers also suggest rebirth, which is the why many people place flowers upon graves.

In ancient times the Greeks would place a crown of flowers on the head of the deceased as well as cover the tomb with wreaths and flower petals, rich and powerful women in ancient Egypt would wear floral headdresses and necklaces upon their death as representations of renewal and a joyful afterlife, and the Mexicans to this day use flowers prominently in their Day of the Dead celebrations in the same way that their Aztec ancestors did.

Fruit

In botany, a fruit is a part of a flowering plant that derives from specific tissues of the flower, mainly one or more ovaries. Fruits are the means by which many plants disseminate seeds. Many plants bearing edible fruits, in particular, have propagated with the movements of humans and animals in a symbiotic relationship as a means for seed dispersal and nutrition, respectively; in fact, humans and many animals have become dependent on fruits as a source of food. Fruits account for a substantial fraction of the world's agricultural output, and some (such as the apple and the pomegranate) have acquired extensive cultural and symbolic meanings.

The section of a fungus that produces spores is also called a fruiting body.

In common language usage, fruit normally means the fleshy seed-associated structures of a plant that are sweet and edible in the raw state, such as apples, oranges, grapes, strawberries, and bananas. On the other hand, the botanical sense includes many structures that are not commonly called "fruits", such as bean pods, corn kernels, wheat grains, and tomatoes.

Botanic Fruit and Culinary Fruit

In the culinary sense of these words, a fruit is usually any sweet-tasting plant product, especially those associated with seeds; a vegetable is any savoury or less sweet plant product; and a nut is any hard, oily, and shelled plant product.

These culinary vegetables that are botanically fruit include cucurbits (e.g., squash, pumpkin, and cucumber), tomatoes, peas, beans, corn, eggplant, and sweet pepper. In addition, some spices, such as allspice and chilies, are fruits, botanically speaking. In contrast, rhubarb is often referred to as a fruit, because it is used to make sweet desserts such as pies, though only the petiole of the rhubarb plant is edible. Edible gymnosperm seeds are often given fruit names, e.g., pine nuts, ginkgo nuts.

Figure: *Euler diagram representing the relationship between (culinary) vegetables and botanical fruits*

Botanically, a cereal grain, such as corn, wheat or rice, is also a kind of fruit, termed a caryopsis. However, the fruit wall is very thin, and is fused to the seed coat, so almost all of the edible grain is actually a seed.

Many common terms for seeds and fruit do not correspond to the botanical classifications. In botany, seeds are ripened ovules; fruits are the ripened ovaries or carpels that contain the seeds and a nut is a type of fruit and not a seed.

Fruit Development

A fruit results from maturation of one or more flowers, and the gynoecium of the flower(s) forms all or part of the fruit.

Inside the ovary/ovaries are one or more ovules where the megagametophyte contains the egg cell. After double fertilization, these ovules will become seeds. The ovules are fertilized in a process that starts with pollination, which involves the movement of pollen from the stamens to the stigma of flowers. After pollination, a tube grows from the pollen through the stigma into the ovary to the ovule and two sperm are transferred from the pollen to the megagametophyte. Within the megagametophyte one of the two sperm unites with the egg, forming a zygote, and the second sperm enters the central cell forming the endosperm mother cell, which completes the double fertilization process. Later the zygote will give rise to the embryo of the seed, and the endosperm mother cell will give rise to endosperm, a nutritive tissue used by the embryo.

As the ovules develop into seeds, the ovary begins to ripen and the ovary wall, the *pericarp*, may become fleshy (as in berries or drupes), or form a hard outer covering (as in nuts). In some multiseeded fruits, the extent to which the flesh develops is proportional to the number of fertilized ovules. The pericarp is often differentiated into two or three distinct layers called the *exocarp* (outer layer, also called epicarp), *mesocarp* (middle layer), and *endocarp* (inner layer). In some fruits, especially simple fruits derived from an inferior ovary, other parts of the flower (such as the floral tube, including the petals, sepals, and stamens), fuse with the ovary and ripen with it. In other cases, the sepals, petals and/or stamens and style of the flower fall off. When such other floral parts are a significant part of the fruit, it is called an *accessory fruit.* Since other parts of the flower may contribute to the structure of the fruit, it is important to study flower structure to understand how a particular fruit forms.

There are three general modes of fruit development:

- Apocarpous fruits develop from a single flower having one or more separate carpels, and they are the simplest fruits.
- Syncarpous fruits develop from a single gynoecium having two or more carpels fused together.
- Multiple fruits form from many different flowers.

Plant scientists have grouped fruits into three main groups, simple fruits, aggregate fruits, and composite or multiple fruits. The groupings are not evolutionarily relevant, since many diverse plant taxa may be in the same group, but reflect how the flower organs are arranged and how the fruits develop.

Aggregate Fruit

Aggregate fruits form from single flowers that have multiple carpels which are not joined together, i.e. each pistil contains one carpel. Each pistil forms a fruitlet, and collectively the fruitlets are called an etaerio. Four types of aggregate fruits include etaerios of achenes, follicles, drupelets, and berries. Ranunculaceae species, including *Clematis* and *Ranunculus* have an etaerio of achenes, *Calotropis* has an etaerio of follicles, and *Rubus* species like raspberry, have an etaerio of drupelets. *Annona* have Etaerio of berries.

The raspberry, whose pistils are termed *drupelets* because each is like a small drupe attached to the receptacle. In some bramble fruits (such as blackberry) the receptacle is elongated and part of the ripe fruit, making the blackberry an *aggregate-accessory* fruit. The strawberry is also an aggregate-accessory fruit, only one in which the seeds are contained in achenes. In all these examples, the fruit develops from a single flower with numerous pistils.

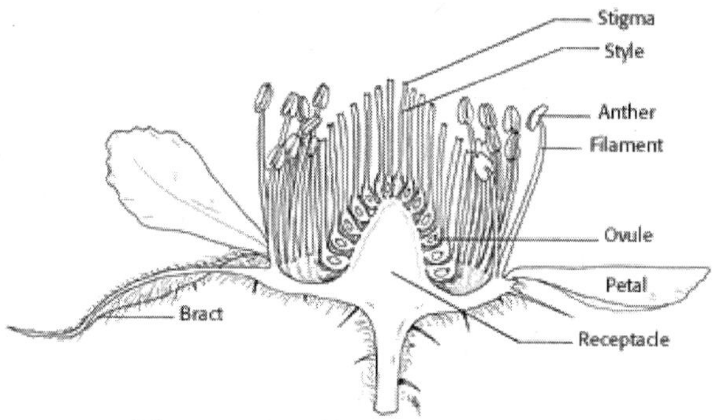

Figure: *Detail of raspberry flower.*

Multiple Fruits

A multiple fruit is one formed from a cluster of flowers (called an *influorescence*). Each flower produces a fruit, but these mature into a single mass. Examples are the pineapple, fig, mulberry, osage-orange, and breadfruit.

In the photograph on the right, stages of flowering and fruit development in the noni or Indian mulberry (*Morinda citrifolia*) can be observed on a single branch. First an influorescence of white flowers called a head is produced. After fertilization, each flower develops into a drupe, and as the drupes expand, they become *connate* (merge) into a *multiple fleshy fruit* called a *syncarpet.*

Accessory Fruit

Some or all of the edible part of accessory fruit is not generated by the ovary.

Seed Dissemination

Variations in fruit structures largely depend on the mode of dispersal of the seeds they contain. This dispersal can be achieved by animals, wind, water, or explosive dehiscence.

Some fruits have coats covered with spikes or hooked burrs, either to prevent themselves from being eaten by animals or to stick to the hairs, feathers or legs of animals, using them as dispersal agents. Examples include cocklebur and unicorn plant.

The sweet flesh of many fruits is "deliberately" appealing to animals, so that the seeds held within are eaten and "unwittingly" carried away and deposited at a distance from the parent. Likewise, the nutritious, oily kernels of nuts are appealing to rodents (such as squirrels) who hoard them in the soil in order to avoid starving during the winter, thus giving those seeds that remain uneaten the chance to germinate and grow into a new plant away from their parent.

Other fruits are elongated and flattened out naturally and so become thin, like wings or helicopter blades, e.g. maple, tuliptree and elm. This is an evolutionary mechanism to increase dispersal distance away from the parent via wind. Other wind-dispersed fruit have tiny *parachutes*, e.g. dandelion and salsify.

Coconut fruits can float thousands of miles in the ocean to spread seeds. Some other fruits that can disperse via water are nipa palm and screw pine.

Some fruits fling seeds substantial distances (up to 100 m in sandbox tree) via explosive dehiscence or other mechanisms, e.g. impatiens and squirting cucumber.

Uses

Many hundreds of fruits, including fleshy fruits like apple, peach, pear, kiwifruit, watermelon and mango are commercially valuable as human food, eaten both fresh and as jams, marmalade and other preserves.

Figure: *Nectarines are one of many fruits that can be easily stewed.*

Fruits are also used in manufactured foods like cookies, muffins, yogurt, ice cream, cakes, and many more. Many fruits are used to make beverages, such as fruit juices (orange juice, apple juice, grape juice, etc.) or alcoholic beverages, such as wine or brandy. Apples are often used to make vinegar. Fruits are also used for gift giving, Fruit Basket and Fruit Bouquet are some common forms of fruit gifts.

Many vegetables are botanical fruits, including tomato, bell pepper, eggplant, okra, squash, pumpkin, green bean, cucumber and zucchini.

Olive fruit is pressed for olive oil. Spices like vanilla, paprika, allspice and black pepper are derived from berries.

Nutritional Value

Fruits are generally high in fibre, water, vitamin C and sugars, although this latter varies widely from traces as in lime, to 61% of the fresh weight of the date. Fruits also contain various phytochemicals that do not yet have an RDA/RDI listing under most nutritional factsheets, and which research indicates are required for proper long-term cellular health and disease prevention. Regular consumption of fruit is associated with reduced risks of cancer, cardiovascular disease (especially coronary heart disease), stroke, Alzheimer disease, cataracts, and some of the functional declines associated with aging.

Diets that include a sufficient amount of potassium from fruits and vegetables also help reduce the chance of developing kidney stones and may help reduce the effects of bone-loss. Fruits are also low in calories which would help lower one's calorie intake as part of a weight-loss diet.

Nonfood Uses

Because fruits have been such a major part of the human diet, different cultures have developed many different uses for various fruits that they do not depend on as being edible. Many dry fruits are used as decorations or in dried flower arrangements, such as unicorn plant, lotus, wheat, annual honesty and milkweed. Ornamental trees and shrubs are often cultivated for their colourful fruits, including holly, pyracantha, viburnum, skimmia, beautyberry and cotoneaster.

Fruits of opium poppy are the source of opium which contains the drugs morphine and codeine, as well as the biologically inactive chemical theabaine from which the drug oxycodone is synthysized. Osage orange fruits are used to repel cockroaches. Bayberry fruits provide a wax often used to make candles. Many fruits provide natural dyes, e.g. walnut, sumac, cherry and mulberry. Dried gourds are used as decorations, water jugs, bird houses, musical instruments, cups and dishes. Pumpkins are carved into Jack-o'-lanterns for Halloween. The spiny fruit of burdock or cocklebur were the inspiration for the invention of Velcro.

Coir is a fibre from the fruit of coconut that is used for doormats, brushes, mattresses, floortiles, sacking, insulation and as a growing medium for container plants. The shell of the coconut fruit is used to make souvenir heads, cups, bowls, musical instruments and bird houses. Fruit is often used as a subject of still life paintings.

Safety

For food safety, the CDC recommends proper fruit handling and preparation to reduce the risk of food contamination and foodborne illness. Fresh fruits and vegetables should be carefully selected. At the store, they should not be damaged or bruised and pre-cut pieces should be refrigerated or surrounded by ice. All fruits and vegetables should be rinsed before eating.

This recommendation also applies to produce with rinds or skins that are not eaten. It should be done just before preparing or eating to avoid premature spoilage. Fruits and vegetables should be kept separate from raw foods like meat, poultry, and seafood, as well as utensils that have come in contact with raw foods. Fruits and vegetables, if they are not going to be cooked, should be thrown away if they have touched raw meat, poultry, seafood or eggs. All cut, peeled, or cooked fruits and vegetables should be refrigerated within two hours. After a certain time, harmful bacteria may grow on them and increase the risk of foodborne illness.

Storage

The plant hormone ethylene causes ripening of many types of fruit. Maintaining most fruits in an efficient cold chain is optimal for post harvest storage, with the aim of extending and ensuring shelf life. All fruits benefit from proper post harvest care.

9

Forest

A forest, also referred to as a wood or the woods, is an area with a high density of trees. As with cities, depending on various cultural definitions, what is considered a forest may vary significantly in size and have different classifications according to how and what of the forest is composed. These plant communities cover approximately 9.4 percent of the Earth's surface (or 30 percent of total land area), though they once covered much more (about 50 percent of total land area), in many different regions and function as habitats for organisms, hydrologic flow modulators, and soil conservers, constituting one of the most important aspects of the biosphere. Although forests are classified primarily by trees, the concept of a forest ecosystem includes additional species (such as smaller plants, fungi, bacteria, and animals) as well as physical and chemical processes such as energy flow and nutrient cycling.

A typical forest is composed of the overstory (canopy or upper tree layer) and the understory. The understory is further subdivided into the shrub layer, herb layer, and also the moss layer and soil microbes. In some complex forests, there is also a well-defined lower tree layer. Forests are central to all human life because they provide a diverse range of resources: they store carbon, aid in regulating the planetary climate, purify water and mitigate natural hazards such as floods. Forests also contain roughly 90 percent of the world's terrestrial biodiversity.

Etymology

The word "forest" comes from Middle English *forest*, from Old French *forest* (also *forès*) "forest, vast expanse covered by trees"; first introduced in English as the word for wild land set aside for hunting without the necessity in definition for the existence of trees (James 1981;Muir 2000,2008). Possibly a borrowing (probably via Frankish or Old High German) of the Medieval Latin word *foresta* “open wood”, *foresta* was first used by Carolingian scribes in the Capitularies of Charlemagne to refer specifically to the king’s royal hunting grounds.

The term was not endemic to Romance languages (e.g. native words for “forest” in the Romance languages evolved out of the Latin word *silva* “forest, wood” (English *sylvan*); cf. Italian, Spanish, Portuguese *selva*; Romanian *silvă*; Old French *selve*); and cognates in Romance languages, such as Italian *foresta*, Spanish and Portuguese *floresta*, etc. are all ultimately borrowings of the French word.

The exact origin of Medieval Latin *foresta* is obscure. Some authorities claim the word derives from the Late Latin phrase *forestam silvam*, meaning “the outer wood”; others claim the term is a latinisation of the Frankish word **forhist* “forest, wooded country”, assimilated to *forestam silvam* (a common practise among Frankish scribes).

Frankish **forhist* is attested by Old High German *forst* “forest”, Middle Low German *vorst* “forest”, Old English *fyrhþ* “forest, woodland, game preserve, hunting ground”, and Old Norse *fýri* “coniferous forest”, all of which derive from Proto-Germanic **fur÷ísa-*, **fur÷íþja-* “a fir-wood, coniferous forest”, from Proto-Indo-European **perkʷu-* “a coniferous or mountain forest, wooded height”.

Figure: *Forest near Rajgir, Bihar, India.*

Uses of the word “forest” in English to denote any uninhabited area of non-enclosure are now considered archaic. The word was introduced by the Norman rulers of England as a legal term (appearing in Latin texts like the Magna Carta) denoting an uncultivated area legally set aside for hunting by feudal nobility. These hunting forests were not necessarily wooded much, if at all. However, as hunting forests did often include considerable areas of woodland, the word "forest" eventually came to mean wooded land more generally. By the start of the fourteenth century the word appeared in English texts, indicating all three senses: the most common one, the legal term and the archaic usage.

Other terms used to mean "an area with a high density of trees" are *wood, woodland, wold, weald, holt, frith* and *firth*. Unlike *forest,* these are all derived from Old English and were not borrowed from another language. Some classifications now reserve the term *woodland* for an area with more open space between trees and distinguish among woodlands, *open forests*, and *closed forests* based on crown cover.

Distribution

Forests can be found in all regions capable of sustaining tree growth, at altitudes up to the tree line, except where natural fire frequency or other disturbance is too high, or where the environment has been altered by human activity.

Figure: *Amazon Rainforest in Brazil*

The latitudes 10° north and south of the Equator are mostly covered in tropical rainforest, and the latitudes between 53°N and 67°N have boreal forest. As a general rule, forests dominated by angiosperms (*broadleaf forests*) are more species-rich than those dominated by gymnosperms (*conifer*, *montane*, or *needleleaf forests*), although exceptions exist.

Forests sometimes contain many tree species only within a small area (as in tropical rain and temperate deciduous forests), or relatively few species over large areas (e.g., taiga and arid montane coniferous forests). Forests are often home to many animal and plant species, and biomass per unit area is high compared to other vegetation communities.

Much of this biomass occurs below ground in the root systems and as partially decomposed plant detritus. The woody component of a forest contains lignin, which is relatively slow to decompose compared with other organic materials such as cellulose or carbohydrate.

Forests are differentiated from woodlands by the extent of canopy coverage: in a forest, the branches and the foliage of separate trees often meet or interlock, although there can be gaps of varying sizes within an area referred to as forest. A woodland has a more continuously open canopy, with trees spaced farther apart, which allows more sunlight to penetrate to the ground between them (also see: savanna).

Among the major forested biomes are:

- rain forest (tropical and temperate)
- taiga
- temperate hardwood forest
- tropical dry forest.

Classification

Forests can be classified in different ways and to different degrees of specificity.

Figure: *Biogradska forest in Montenegro.*

One such way is in terms of the "biome" in which they exist, combined with leaf longevity of the dominant species (whether they are evergreen or deciduous). Another distinction is whether the forests are composed predominantly of broadleaf trees, coniferous (needle-leaved) trees, or mixed.

- Boreal forests occupy the subarctic zone and are generally evergreen and coniferous.
- Temperate zones support both broadleaf deciduous forests (*e.g.*, temperate deciduous forest) and evergreen coniferous forests (*e.g.*, temperate coniferous forests and temperate rainforests). Warm temperate zones support broadleaf evergreen forests, including laurel forests.
- Tropical and subtropical forests include tropical and subtropical moist forests, tropical and subtropical dry forests, and tropical and subtropical coniferous forests.
- Physiognomy classifies forests based on their overall physical structure or developmental stage (e.g. old growth vs. second growth).
- Forests can also be classified more specifically based on the climate and the dominant tree species present, resulting in numerous different forest types (e.g., ponderosa pine/Douglas-fir forest).

A number of global forest classification systems have been proposed, but none has gained universal acceptance. UNEP-WCMC's forest category classification system is a simplification of other more complex systems (e.g. UNESCO's forest and woodland 'subformations'). This system divides the world's forests into 26 major types, which reflect climatic zones as well as the principal types of trees. These 26 major types can be reclassified into 6 broader categories: temperate needleleaf; temperate broadleaf and mixed; tropical moist; tropical dry; sparse trees and parkland; and forest plantations. Each category is described as a separate section below.

Temperate Needleleaf

Temperate needleleaf forests mostly occupy the higher latitude regions of the northern hemisphere, as well as high altitude zones and some warm temperate areas, especially on nutrient-poor or otherwise unfavourable soils. These forests are composed entirely, or nearly so, of coniferous species (Coniferophyta). In the Northern Hemisphere pines Pinus, spruces Picea, larches Larix, silver firs Abies, Douglas firs Pseudotsuga and hemlocks Tsuga, make up the canopy, but other taxa are also important. In the Southern Hemisphere, most coniferous trees (members of the Araucariaceae and Podocarpaceae) occur in mixtures with broadleaf species that are classed as broadleaf and mixed forests.

Temperate Broadleaf and Mixed

Temperate broadleaf and mixed forests include a substantial component of trees in the Anthophyta. They are generally characteristic of the warmer temperate latitudes, but extend to cool temperate ones, particularly in the southern hemisphere. They include such forest types as the mixed deciduous forests of the United States and their counterparts in China and Japan, the broadleaf evergreen rainforests of Japan, Chile and Tasmania, the sclerophyllous forests of Australia, central Chile, the Mediterranean and California, and the southern beech Nothofagus forests of Chile and New Zealand.

Tropical Moist

There are many different types of tropical moist forests,although most extensive are the lowland evergreen broad leaf rainforests, for example várzea and igapó forests and the terra firma forests of the Amazon Basin; the peat swamp forests, dipterocarp forests of Southeast Asia; and the high forests of the Congo Basin. Forests located on mountains are also included in this category, divided largely into upper and lower montane formations on the basis of the variation of physiognomy corresponding to changes in altitude.

Tropical Dry

Tropical dry forests are characteristic of areas in the tropics affected by seasonal drought. The seasonality of rainfall is usually reflected in the deciduousness of the forest canopy, with most trees being leafless for several months of the year. However, under some conditions, e.g. less fertile soils or less predictable drought regimes, the proportion of evergreen species increases and the forests are characterised as "sclerophyllous".

Thorn forest, a dense forest of low stature with a high frequency of thorny or spiny species, is found where drought is prolonged, and especially where grazing animals are plentiful. On very poor soils, and especially where fire is a recurrent phenomenon, woody savannas develop.

Sparse Trees and Parkland

Sparse trees and parkland are forests with open canopies of 10–30% crown cover. They occur principally in areas of transition from forested to non-forested landscapes. The two major zones in which these ecosystems occur are in the boreal region and in the seasonally dry tropics. At high latitudes, north of the main zone of boreal forest or taiga, growing conditions are not adequate to maintain a continuous closed forest cover, so tree cover is both sparse and discontinuous. This vegetation is variously called open taiga, open lichen woodland, and forest tundra. It is species-poor, has high bryophyte cover, and is frequently affected by fire.

Forest Plantations

Forest plantations, generally intended for the production of timber and pulpwood increase the total area of forest worldwide. Commonly mono-specific and/or composed of introduced tree species, these ecosystems are not generally important as habitat for native biodiversity. However, they can be managed in ways that enhance their biodiversity protection functions and they are important providers of ecosystem services such as maintaining nutrient capital, protecting watersheds and soil structure as well as storing carbon. They may also play an important role in alleviating pressure on natural forests for timber and fuelwood production.

Forest Categories

28 forest categories are used to enable the translation of forest types from national and regional classification systems to a harmonised global one:

Figure: *A temperate deciduous broadleaf forest, the* Hasenholz, *southeast of Kirchheim unter Teck, Baden-Württemberg, Germany*

Temperate and Boreal Forest Types

1. Evergreen needleleaf forest – Natural forest with > 30% canopy cover, in which the canopy is predominantly (> 75%) needleleaf and evergreen.
2. Deciduous needleleaf forests – Natural forests with > 30% canopy cover, in which the canopy is predominantly (> 75%) needleleaf and deciduous.
3. Mixed broadleaf/needleleaf forest – Natural forest with > 30% canopy cover, in which the canopy is composed of a more or less even mixture of needleleaf and broadleaf crowns (between 50:50% and 25:75%).
4. Broadleaf evergreen forest – Natural forests with > 30% canopy cover, the canopy being > 75% evergreen and broadleaf.
5. Deciduous broadleaf forest – Natural forests with > 30% canopy cover, in which > 75% of the canopy is deciduous and broadleaves predominate (> 75% of canopy cover).
6. Freshwater swamp forest – Natural forests with > 30% canopy cover, composed of trees with any mixture of leaf type and seasonality, but in which the predominant environmental characteristic is a waterlogged soil.
7. Sclerophyllous dry forest – Natural forest with > 30% canopy cover, in which the canopy is mainly composed of sclerophyllous broadleaves and is > 75% evergreen.
8. Disturbed natural forest – Any forest type above that has in its interior significant areas of disturbance by people, including clearing, felling for wood extraction, anthropogenic fires, road construction, etc.
9. Sparse trees and parkland – Natural forests in which the tree canopy cover is between 10–30%, such as in the steppe regions of the world. Trees of any type (e.g., needleleaf, broadleaf, palms).
10. Exotic species plantation – Intensively managed forests with > 30% canopy cover, which have been planted by people with species not naturally occurring in that country.
11. Native species plantation – Intensively managed forests with > 30% canopy cover, which have been planted by people with species that occur naturally in that country.
12. *Unspecified forest plantation – Forest plantations showing extent only with no further information about their type, This data currently only refers to the Ukraine.

13. *Unclassified forest data – Forest data showing forest extent only with no further information about their type.

Those marked * have been created as a result of data holdings which do not specify the forest type, hence 26 categories are quoted, not 28 shown here.

Tropical Forest Types

1. Lowland evergreen broadleaf rain forest – Natural forests with > 30% canopy cover, below 1,200 m (3,937 ft) altitude that display little or no seasonality, the canopy being >75% evergreen broadleaf.
2. Lower montane forest – Natural forests with > 30% canopy cover, between 1200–1800 m altitude, with any seasonality regime and leaf type mixture.
3. Upper montane forest – Natural forests with > 30% canopy cover, above 1,800 m (5,906 ft) altitude, with any seasonality regime and leaf type mixture.
4. Freshwater swamp forest – Natural forests with > 30% canopy cover, below 1,200 m (3,937 ft) altitude, composed of trees with any mixture of leaf type and seasonality, but in which the predominant environmental characteristic is a waterlogged soil.

Figure: *The Fatu Hiva rainforest in Polynesia.*

5. Semi-evergreen moist broadleaf forest – Natural forests with > 30% canopy cover, below 1,200 m (3,937 ft) altitude in which between 50–75% of the canopy is evergreen, > 75% are broadleaves, and the trees display seasonality of flowering and fruiting.

6. Mixed broadleaf/needleleaf forest – Natural forests with > 30% canopy cover, below 1,200 m (3,937 ft) altitude, in which the canopy is composed of a more or less even mixture of needleleaf and broadleaf crowns (between 50:50% and 25:75%).
7. Needleleaf forest – Natural forest with > 30% canopy cover, below 1,200 m (3,937 ft) altitude, in which the canopy is predominantly (> 75%) needleleaf.
8. Mangroves – Natural forests with > 30% canopy cover, composed of species of mangrove tree, generally along coasts in or near brackish or seawater.
9. Disturbed natural forest – Any forest type above that has in its interior significant areas of disturbance by people, including clearing, felling for wood extraction, anthropogenic fires, road construction, etc.
10. Deciduous/semi-deciduous broadleaf forest – Natural forests with > 30% canopy cover, below 1,200 m (3,937 ft) altitude in which between 50–100% of the canopy is deciduous and broadleaves predominate (> 75% of canopy cover).
11. Sclerophyllous dry forest – Natural forests with > 30% canopy cover, below 1,200 m (3,937 ft) altitude, in which the canopy is mainly composed of sclerophyllous broadleaves and is > 75% evergreen.
12. Thorn forest – Natural forests with > 30% canopy cover, below 1,200 m (3,937 ft) altitude, in which the canopy is mainly composed of deciduous trees with thorns and succulent phanerophytes with thorns may be frequent.
13. Sparse trees and parkland – Natural forests in which the tree canopy cover is between 10–30%, such as in the savannah regions of the world. Trees of any type (e.g., needleleaf, broadleaf, palms).
14. Exotic species plantation – Intensively managed forests with > 30% canopy cover, which have been planted by people with species not naturally occurring in that country.
15. Native species plantation – Intensively managed forests with > 30% canopy cover, which have been planted by people with species that occur naturally in that country.

Forest Loss and Management

The scientific study of forest species and their interaction with the environment is referred to as forest ecology, while the management of forests is often referred to as forestry. Forest management has changed considerably over the last few centuries, with rapid changes from the 1980s onwards culminating in a practice now referred to as sustainable forest management. Forest ecologists concentrate on forest patterns and processes, usually with the aim of elucidating cause and effect relationships. Foresters who practice sustainable forest management focus on the integration of ecological, social and economic values, often in consultation with local communities and other stakeholders.

Anthropogenic factors that can affect forests include logging, urban sprawl, human-caused forest fires, acid rain, invasive species, and the slash and burn practices of swidden agriculture or shifting cultivation. The loss and re-growth of forest leads to a distinction between two broad types of forest, primary or old-growth forest and secondary forest.

There are also many natural factors that can cause changes in forests over time including forest fires, insects, diseases, weather, competition between species, etc. In 1997, the World Resources Institute recorded that only 20% of the world's original forests remained in large intact tracts of undisturbed forest. More than 75% of these intact forests lie in three countries – the Boreal forests of Russia and Canada and the rainforest of Brazil. In 2006 this information on intact forests was updated using latest available satellite imagery.

Canada has about 4,020,000 square kilometres (1,550,000 sq mi) of forest land. More than 90% of forest land is publicly owned and about 50% of the total forest area is allocated for harvesting. These allocated areas are managed using the principles of sustainable forest management, which includes extensive consultation with local stakeholders.

About eight percent of Canada's forest is legally protected from resource development (Global Forest Watch Canada)(Natural Resources Canada). Much more forest land – about 40 percent of the total forest land base – is subject to varying degrees of protection through processes such as integrated land use planning or defined management areas such as certified forests (Natural Resources Canada).

These maps represent only virgin forest lost. Some regrowth has occurred but not to the age, size or extent of 1620 due to population increases and food cultivation. *From William B. Greeley's, The Relation of Geography to Timber Supply, Economic Geography, 1925, vol. 1, p. 1–11. Source of "Today" map: compiled by George Draffan from roadless area map in The Big Outside: A Descriptive Inventory of the Big Wilderness Areas of the United States, by Dave Foreman and Howie Wolke (Harmony Books, 1992).'*

By December 2006, over 1,237,000 square kilometers of forest land in Canada (about half the global total) had been certified as being sustainably managed (Canadian Sustainable Forestry Certification Coalition). Clearcutting, first used in the latter half of the 20th century, is less expensive, but devastating to the environment and companies are required by law to ensure that harvested areas are adequately regenerated.

Most Canadian provinces have regulations limiting the size of clearcuts, although some older clearcuts can range upwards of 110 square kilometres (27,000 acres) in size which were cut over several years. China instituted a ban on logging, beginning in 1998, due to the destruction caused by clearcutting. Selective cutting avoids the erosion, and flooding, that result from clearcutting.

In the United States, most forests have historically been affected by humans to some degree, though in recent years improved forestry practices has helped regulate or moderate large scale or severe impacts. However, the United States Forest Service estimates a net loss of about 2 million hectares (4,942,000 acres) between 1997 and 2020; this estimate includes conversion of forest land to other uses, including urban and suburban development, as well as afforestation and natural reversion of abandoned crop and pasture land to forest.

However, in many areas of the United States, the area of forest is stable or increasing, particularly in many northern states. The opposite problem from flooding has plagued national forests, with loggers complaining that a lack of thinning and proper forest management has resulted in large forest fires.

Old-growth forest contains mainly natural patterns of biodiversity in established seral patterns, and they contain mainly species native to the region and habitat. The natural formations and processes have not been affected by humans with a frequency or intensity to change the natural structure and components of the habitat. Secondary forest contains significant elements of species which were originally from other regions or habitats.

Smaller areas of woodland in cities may be managed as Urban forestry, sometimes within public parks. These are often created for human benefits; Attention Restoration Theory argues that spending time in nature reduces stress and improves health, while forest schools and kindergartens help young people to develop social as well as scientific skills in forests. These typically need to be close to where the children live, for practical logistics.

Plant Cell

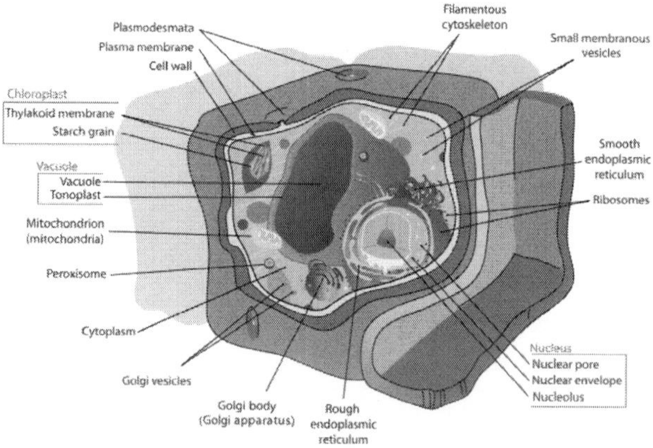

Figure: *Plant cell structure*

Plant cells are eukaryotic cells that differ in several key respects from the cells of other eukaryotic organisms. Their distinctive features include:

- A large central vacuole, a water-filled volume enclosed by a membrane known as the *tonoplast* maintains the cell's turgor, controls movement of molecules between the cytosol and sap, stores useful material and digests waste proteins and organelles.
- A cell wall composed of cellulose and hemicellulose, pectin and in many cases lignin, is secreted by the protoplast on the outside of the cell membrane. This contrasts with the cell walls of fungi (which are made of chitin), and of bacteria, which are made of peptidoglycan.
- Specialised cell–cell communication pathways known as plasmodesmata, pores in the primary cell wall through which the plasmalemma and endoplasmic reticulum of adjacent cells are continuous.
- Plastids, the most notable being the chloroplasts, which contain chlorophyll a green coloured pigment which is used for absorbing sunlight and is used by a plant to make its own food in the process is known as photosynthesis. Other types of plastid are the amyloplasts, specialized for starch storage, elaioplasts specialized for fat storage, and chromoplasts specialized for

synthesis and storage of pigments. As in mitochondria, which have a genome encoding 37 genes, plastids have their own genomes of about 100–120 unique genes and, it is presumed, arose as prokaryotic endosymbionts living in the cells of an early eukaryotic ancestor of the land plants and algae.

- Cell division by construction of a phragmoplast as a template for building a cell plate late in cytokinesis is characteristic of land plants and a few groups of algae, notably the Charophytes and the Order Trentepohliales
- The sperm of bryophytes and pteridophytes have flagellae similar to those in animals, but higher plants, (including Gymnosperms and flowering plants) lack the flagellae and centrioles that are present in animal cells.

Cell Types

- Parenchyma cells are living cells that have really diverse functions ranging from storage and support to photosynthesis and phloem loading (transfer cells). Apart from the xylem and phloem in their vascular bundles, leaves are composed mainly of parenchyma cells. Some parenchyma cells, as in the epidermis, are specialized for light penetration and focusing or regulation of gas exchange, but others are among the least specialized cells in plant tissue, and may remain totipotent, capable of dividing to produce new populations of undifferentiated cells, throughout their lives. Parenchyma cells have thin, permeable primary walls enabling the transport of small molecules between them, and their cytoplasm is responsible for a wide range of biochemical functions such as nectar secretion, or the manufacture of secondary products that discourage herbivory. Parenchyma cells that contain many chloroplasts and are concerned primarily with photosynthesis are called chlorenchyma cells. Others, such as the majority of the parenchyma cells in potato tubers and the seed cotyledons of legumes, have a storage function.
- Collenchyma cells – collenchyma cells are alive at maturity and have only a primary wall. These cells mature from meristem derivatives that initially resemble parenchyma, but differences quickly become apparent. Plastids do not develop, and the secretory apparatus (ER and Golgi) proliferates to secrete additional primary wall. The wall is most commonly thickest at the corners, where three or more cells come in contact, and

thinnest where only two cells come in contact, though other arrangements of the wall thickening are possible.

Pectin and hemicellulose are the dominant constituents of collenchyma cell walls of dicotyledon angiosperms, which may contain as little as 20% of cellulose in *Petasites*. Collenchyma cells are typically quite elongated, and may divide transversely to give a septate appearance. The role of this cell type is to support the plant in axes still growing in length, and to confer flexibility and tensile strength on tissues. The primary wall lacks lignin that would make it tough and rigid, so this cell type provides what could be called plastic support – support that can hold a young stem or petiole into the air, but in cells that can be stretched as the cells around them elongate. Stretchable support (without elastic snap-back) is a good way to describe what collenchyma does. Parts of the strings in celery are collenchyma.

- Sclerenchyma cells – Sclerenchyma cells (from the Greek skleros, *hard*) are hard and tough cells with a function in mechanical support. They are of two broad types – sclereids or stone cells and fibres. The cells develop an extensive secondary cell wall that is laid down on the inside of the primary cell wall. The secondary wall is impregnated with lignin, making it hard and impermeable to water. Thus, these cells cannot survive for long' as they cannot exchange sufficient material to have active metabolism. Sclerenchyma cells are typically dead at functional maturity, and the cytoplasm is missing, leaving an empty central cavity.

Functions for sclereid cells (hard cells that give leaves or fruits a gritty texture) include discouraging herbivory, by damaging digestive passages in small insect larval stages, and physical protection (a solid tissue of hard sclereid cells form the pit wall in a peach and many other fruits). Functions of fibres include provision of load-bearing support and tensile strength to the leaves and stems of herbaceous plants. Sclerenchyma fibres are not involved in conduction, either of water and nutrients (as in the xylem) or of carbon compounds (as in the phloem), but it is likely that they may have evolved as modifications of xylem and phloem initials in early land plants.

Tissue Types

The major classes of cells differentiate from undifferentiated meristematic cells (analogous to the stem cells of animals) to form the tissue structures of roots, stems, leaves, flowers, and reproductive structures.

Xylem cells are elongated cells with lignified secondary thickening of the cell walls. Xylem cells are specialised for conduction of water, and first appeared in plants during their transition to land in the Silurian period more than 425 million years ago. The possession of xylem defines the vascular plants or Tracheophytes. Xylem tracheids are pointed, elongated xylem cells, the simplest of which have continuous primary cell walls and lignified secondary wall thickenings in the form of rings, hoops, or reticulate networks. More complex tracheids with valve-like perforations called bordered pits characterise the gymnosperms. The ferns and other pteridophytes and the gymnosperms have only xylem tracheids, while the angiosperms also have xylem vessels. Vessel members are hollow xylem cells aligned end-to-end, without end walls that are assembled into long continuous tubes. The bryophytes lack true xylem cells, but their sporophytes have a water-conducting tissue known as the hydrome that is composed of elongated cells of simpler construction.

Phloem is a specialised tissue for food conduction in higher plants. The conduction of food is a complex process that is carried in the plant with the help of special cell called phloem cells. These cells conduct inter- and intra-cellular fluid (food – proteins and other essential elements required by the plant for its metabolism) through the process of osmosis. This phenomenon is called ascent of sap in plants. Phloem consists of two cell types, the sieve tubes and the intimately-associated companion cells. The sieve tube elements lack nuclei and ribosomes, and their metabolism and functions are regulated by the adjacent nucleate companion cells. Sieve tubes are joined end-to-end with perforate end-plates between known as *sieve plates*, which allow transport of photosynthate between the sieve elements. The companion cells, connected to the sieve tubes via plasmodesmata, are responsible for loading the phloem with sugars. The bryophytes lack phloem, but moss sporophytes have a simpler tissue with analogous function known as the leptome.

Plant epidermal cells are specialised parenchyma cells covering the external surfaces of leaves, stems and roots. The epidermal cells of aerial organs arise from the superficial layer of cells known as the *tunica* (L1 and L2 layers) that covers the plant shoot apex, whereas the cortex and vascular tissues arise from innermost layer of the shoot apex known as the *corpus* (L3 layer). The epidermis of roots originates from the layer of cells immediately beneath the root cap.

The epidermis of all aerial organs, but not roots, is covered with a cuticle made of the polyester cutin with a superficial layer of waxes.

The epidermal cells of the primary shoot are thought to be the only plant cells with the biochemical capacity to synthesize cutin. Several cell types may be present in the epidermis. Notable among these are the stomatal guard cells, glandular and clothing hairs or trichomes, and the root hairs of primary roots. In the shoot epidermis of most plants, only with really good chloroplasts.

Pollen

Pollen is a fine to coarse powder containing the microgametophytes of seed plants, which produce the male gametes (sperm cells). Pollen grains have a hard coat that protects the sperm cells during the process of their movement from the stamens to the pistil of flowering plants or from the male cone to the female cone of coniferous plants. When pollen lands on a compatible pistil or female cone (i.e., when pollination has occurred), it germinates and produces a pollen tube that transfers the sperm to the ovule (or female gametophyte). Individual pollen grains are small enough to require magnification to see detail. The study of pollen is called palynology and is highly useful in paleoecology, paleontology, archeology, and forensics.

The Structure and Formation of Pollen

Pollen itself is not the male gamete. Each pollen grain contains vegetative (non-reproductive) cells (only a single cell in most flowering plants but several in other seed plants) and a generative (reproductive) cell containing two nuclei: a tube nucleus (that produces the pollen tube) and a generative nucleus (that divides to form the two sperm cells). The group of cells is surrounded by a cellulose-rich cell wall called the intine, and a resistant outer wall composed largely of sporopollenin called the exine.

Pollen is produced in the 'microsporangium' (contained in the anther of an angiosperm flower, male cone of a coniferous plant, or male cone of other seed plants). Pollen grains come in a wide variety of shapes (most often spherical), sizes, and surface markings characteristic of the species. Pollen grains of pines, firs, and spruces are winged. The smallest pollen grain, that of the forget-me-not (*Myosotis* spp.), is around 6 μm (0.006 mm) in diameter. Wind-borne pollen grains can be as large as about 90-100 μm.

In angiosperms, during flower development the anther is composed of a mass of cells that appear undifferentiated, except for a partially differentiated dermis. As the flower develops, four groups of sporogenous cells form within the anther. The fertile sporogenous cells are surrounded by layers of sterile cells that grow into the wall of the pollen sac. Some of the cells grow into nutritive cells that supply nutrition for the microspores that form by meiotic division from the sporogenous cells. In a process called microsporogenesis, four haploid microspores are produced from each diploid sporogenous cell (microsporocyte), after meiotic division. After the formation of the four microspores, which are contained by callose walls, the development of the pollen grain walls begins. The callose wall is broken down by an enzyme called callase and the freed pollen grains grow in size and develop their characteristic shape and form a resistant outer wall called the exine and an inner wall called the intine. The exine is what is preserved in the fossil record.

The pollen wall protects the sperm while the pollen grain is moving from the anther to the stigma; it protects the vital genetic material from drying out and solar radiation. The pollen grain surface is covered with waxes and proteins, which are held in place by structures called sculpture elements on the surface of the grain. The outer pollen wall, which prevents the pollen grain from shrinking and crushing the genetic material during desiccation, is composed of two layers.

These two layers are the tectum and the foot layer, which is just above the intine. The tectum and foot layer are separated by a region called the columella, which is composed of strengthening rods. The outer wall is constructed with a resistant biopolymer called sporopollenin. The pollen tube passes through the wall by way of structures called apertures.

Pollen apertures are various modifications of the wall of the pollen grain that may involve thinning, ridges and pores. They serve as an exit for the pollen contents and allow shrinking and swelling of the grain caused by changes in moisture content. The elongated apertures/ furrows in the pollen grain are called colpi (singular: colpus) which along with pores, are a major criterion for the identification of classes of pollen.

The orientation of furrows (relative to the original tetrad of microspores) classify the pollen as colpate or sulcate. Eudicots have three colpi (tricolpate) or pollen shapes that are evolutionarily derived from tricolpate pollen. Other groups having one sulcus (monosulcate), monoc.

Except in the case of some submerged aquatic plants, the mature pollen-grain has a double wall, a thin delicate wall of unaltered cellulose (the endospore or intine) and a tough outer cuticularized exospore or exine. The exine often bears spines or warts, or is variously sculptured, and the character of the markings is often of value for identifying genus, species, or even cultivar or individual. In some flowering plants, germination of the pollen grain often begins before it leaves the microsporangium, with the generative cell forming the two sperm cells.

Pollination

The transfer of pollen grains to the female reproductive structure (*pistil* in angiosperms) is called pollination. This transfer can be mediated by the wind, in which case the plant is described as anemophilous (literally wind-loving). Anemophilous plants typically produce great quantities of very lightweight pollen grains, sometimes with air-sacs. Non-flowering seed plants (e.g. pine trees) are characteristically anemophilous.

Anemophilous flowering plants generally have inconspicuous flowers. Entomophilous (literally insect-loving) plants produce pollen that is relatively heavy, sticky and protein-rich, for dispersal by insect pollinators attracted to their flowers. Many insects and some mites are specialized to feed on pollen, and are called palynivores.

In non-flowering seed plants, pollen germinates in the pollen chamber, located beneath and inside the micropyle. A pollen tube is produced, which grows into the nucellus to provide nutrients for the developing sperm cells. Sperm cells of Pinophyta and Gnetophyta are without flagella, and are carried by the pollen tube, while those of Cycadophyta and Ginkgophyta have many flagella.

When placed on the stigma of a flowering plant, under favourable circumstances, a pollen grain puts forth a pollen tube which grows down the tissue of the style to the ovary, and makes its way along the placenta, guided by projections or hairs, to the micropyle of an ovule. The nucleus of the tube cell has meanwhile passed into the tube, as does also the generative nucleus which divides (if it hasn't already) to form two sperm cells. The sperm cells are carried to their destination in the tip of the pollen-tube.

Pollen as a Carrier of Ecological Information in Plants

A Russian theoretical biologist has suggested that the quantity of pollen reaching a pistillate flower can transmit ecological information and also regulate evolutionary plasticity in cross-pollinating plants. Plentiful pollen indicates optimum environmental conditions (for example a plant that is situated at the centre of its natural range, in ideal growing conditions, with a large number of male plants nearby, and favourable weather conditions), whereas a small amount of pollen indicates extreme conditions (at the borders of its range, with a deficiency of male plants, and adverse weather conditions).

Geodakian believes that the quantity of pollen reaching a pistillate flower defines the sex ratio, dispersion and sexual dimorphism of a plant population. High pollen quantity leads to a reduction of these characteristics and stabilization of a population. Small quantity leads to their increase and destabilization of a population.

Dependence of the secondary sex ratio on the amount of fertilizing pollen was confirmed on four dioecious plant species from three families — Rumex acetosa (Polygonaceae), Melandrium album (Cariophyllaceae), Cannabis sativa and Humulus japonicus (Cannabinaceae).

Dependence of offspring phenotype variety on amount of pollen was observed by Ter-Avanesyan in 1949. All three studied species of plants (cotton plant, black-eyed pea, and wheat) showed dependence in the direction forecast by the theory — fertilization with a small amount of pollen resulted in an increase in the diversity of the offspring. Ter-Avanesian writes that as a result of a limited pollination "instead of homogenous sorts we get populations".

Pollen in the Fossil Record

Pollen's sporopollenin outer sheath affords it some resistance to the rigours of the fossilisation process that destroy weaker objects; it is also produced in huge quantities. As such, there is an extensive fossil record of pollen grains, often disassociated from their parent plant. The discipline of palynology is devoted to the study of pollen, which can be used both for biostratigraphy and to gain information about the abundance and variety of plants alive — which can itself yield important information about paleoclimates. Pollen is first found in the fossil record in the late Devonian period and increases in abundance until the present day.

Hay Fever

Allergy to pollen is called hay fever. Generally pollens that cause allergies are those of anemophilous plants (pollen is dispersed by air currents.) Such plants produce large quantities of lightweight pollen (because wind dispersal is random and the likelihood of one pollen grain landing on another flower is small) which can be carried for great distances and are easily inhaled, bringing it into contact with the sensitive nasal passages.

In the US, people often mistakenly blame the conspicuous goldenrod flower for allergies. Since this plant is entomophilous (its pollen is dispersed by animals), its heavy, sticky pollen does not become independently airborne. Most late summer and fall pollen allergies are probably caused by ragweed, a widespread anemophilous plant.

Arizona was once regarded as a haven for people with pollen allergies, although several ragweed species grow in the desert. However, as suburbs grew and people began establishing irrigated lawns and gardens, more irritating species of ragweed gained a foothold and Arizona lost its claim of freedom from hay fever.

Anemophilous spring blooming plants such as oak, birch, hickory, pecan, and early summer grasses may also induce pollen allergies. Most cultivated plants with showy flowers are entomophilous and do not cause pollen allergies. In the US, oak pollen starts to cause problems for sufferers in February and is gone by the end of April.

The percentage of people in the United States affected by hay fever varies between 10% and 20%, and such allergy has proven to be the most frequent allergic response in the nation. There are certain evidential suggestions pointing out hay fever and similar allergies to be of hereditary origin. Individuals who suffer from eczema or are asthmatic tend to be more susceptible to developing long-term hay fever.

The most efficient way to handle a pollen allergy is by preventing contact with the material. Individuals carrying the ailment may at first believe that they have a simple summer cold, but hay fever becomes more evident when the apparent cold does not disappear. The confirmation of hay fever can be obtained after examination by a general physician.

Treatment

Antihistamines are effective at treating mild cases of hay fever, this type of non-prescribed drugs includes loratadine, cetirizine and chlorphenamine. They do not prevent the discharge of histamine, but it has been proven that they do prevent a part of the chain reaction activated by this biogenic amine, which considerably lowers hay fever symptoms. A side effect of antihistamines is somnolence, and it is therefore recommended not to take these drugs while driving an automobile or during the consumption of alcoholic beverages. However, the side effects of these medications can vary from person to person.

Decongestants can be administered in different ways such as tablets and nasal sprays. Decongestants such as pseudoephedrine, xylometazoline and drixoral can be acquired as over-the-counter medications. Since oral decongestant drugs raise blood pressure levels, individuals with hypertension are advised to avoid them. The oral decongestant type can aggravate the symptoms of an enlarged prostate, making the process of urinating more complicated.

Nutrition

Most major classes of predatory and parasitic arthropods contain species that eat pollen, despite the common perception that bees are the primary pollen-consuming arthropod group. Many other Hymenoptera other than bees consume pollen as adults, though only a small number feed on pollen as larvae (including some ant larvae). Spiders are normally considered carnivores but pollen is an important source of food for several species, particularly for spiderlings which catch pollen on their webs.

It is not clear how spiderlings manage to eat pollen however, since their mouths are not large enough to consume pollen grains. Some predatory mites also feed on pollen, with some species being able to subsist solely on pollen, such as *Euseius tularensis*, which feeds on the pollen of dozens of plant species. Members of some beetle families such as Mordellidae and Melyridae feed almost exclusively on pollen as adults, while various lineages within larger families such as Curculionidae, Chrysomelidae, Cerambycidae, and Scarabaeidae are pollen specialists even though most members of their families are not (e.g., only 36 of 40000 species of ground beetles, which are typically predatory, have been shown to eat pollen — but this is thought to be a severe underestimate as the feeding habits are only known for 1000 species). Similarly, Ladybird beetles mainly eat insects, but many species also eat pollen, as either part or all of their diet. Hemiptera are mostly herbivores or omnivores but pollen feeding is known (and has only been well studied in the Anthocoridae). Many adult flies, especially Syrphidae, feed on pollen, and three UK syrphid species feed strictly on pollen (syrphids, like all flies, cannot eat pollen directly due to the structure of their mouthparts, but can consume pollen contents that are dissolved in a fluid). Some species of fungus, including *Fomes fomentarius*, are able to break down grains of pollen as a secondary nutrition source which is particularly high in nitrogen.

Although bats, butterflies and hummingbirds are not pollen eaters *per se*, their consumption of nectar in flowers is an important aspect of the pollination process.

In Humans

A variety of producers have started selling bee pollen for human consumption, often marketed as a food (rather than a dietary supplement). The largest constituent is carbohydrates, with protein content ranging from 7 to 35 percent depending on the plant species collected by bees.

The U.S. Food and Drug Administration (FDA) has not found any harmful effects of pollen consumption, except from the usual allergies. However, FDA does not allow pollen marketers in the United States to make health claims about their produce, as no scientific basis for these has ever been proven. Furthermore, there are possible dangers not only from allergic reactions but also from contaminants such as pesticides and from fungi and bacteria growth related to poor storage procedures. A manufacturers's claim that pollen collecting helps the bee colonies is also controversial.

Forensic Palynology

In forensic biology, pollen can tell a lot about where a person or object has been, because regions of the world, or even more particular locations such a certain set of bushes, will have a distinctive collection of pollen species. Pollen evidence can also reveal the season in which a particular object picked up the pollen. Pollen has been used to trace activity at mass graves in Bosnia, catch a burglar who brushed against a *Hypericum* bush during a crime, and has even been proposed as an additive for bullets to enable tracking them.

Seed

A seed is a small embryonic plant enclosed in a covering called the seed coat, usually with some stored food. It is the product of the ripened ovule of gymnosperm and angiosperm plants which occurs after fertilization and some growth within the mother plant. The formation of the seed completes the process of reproduction in seed plants (started with the development of flowers and pollination), with the embryo developed from the zygote and the seed coat from the integuments of the ovule.

Seeds have been an important development in the reproduction and spread of flowering plants, relative to more primitive plants such as mosses, ferns and liverworts, which do not have seeds and use other means to propagate themselves. This can be seen by the success of seed plants (both gymnosperms and angiosperms) in dominating biological niches on land, from forests to grasslands both in hot and cold climates. The term "seed" also has a general meaning that antedates the above — anything that can be sown, e.g. "seed" potatoes, "seeds" of corn or sunflower "seeds". In the case of sunflower and corn "seeds", what is sown is the seed enclosed in a shell or husk, whereas the potato is a tuber.

Seed Structure

A typical seed includes three basic parts: (1) an embryo, (2) a supply of nutrients for the embryo, and (3) a seed coat.

The embryo is an immature plant from which a new plant will grow under proper conditions. The embryo has one cotyledon or seed leaf in monocotyledons, two cotyledons in almost all dicotyledons and two or more in gymnosperms. The radicle is the embryonic root. The plumule is the embryonic shoot. The embryonic stem above the point of attachment of the cotyledon(s) is the epicotyl. The embryonic stem below the point of attachment is the hypocotyl.

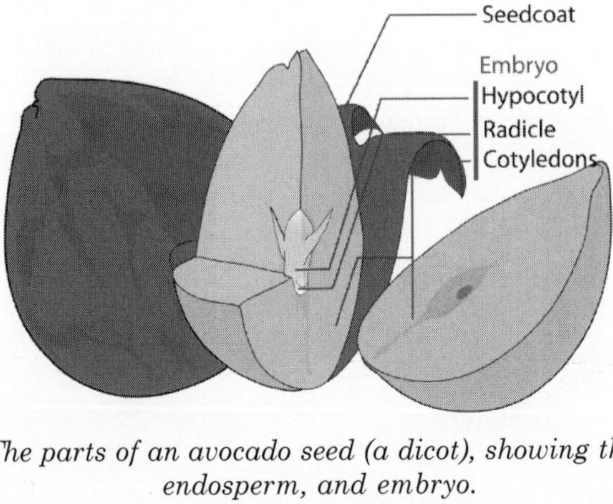

Figure: *The parts of an avocado seed (a dicot), showing the seed coat, endosperm, and embryo.*

Within the seed, there usually is a store of nutrients for the seedling that will grow from the embryo. The form of the stored nutrition varies depending on the kind of plant. In angiosperms, the stored food begins as a tissue called the endosperm, which is derived from the parent plant via double fertilization. The usually triploid endosperm is rich in oil or starch and protein.

In gymnosperms, such as conifers, the food storage tissue (also called endosperm) is part of the female gametophyte, a haploid tissue. In some species, the embryo is embedded in the endosperm or female gametophyte, which the seedling will use upon germination. In others, the endosperm is absorbed by the embryo as the latter grows within the developing seed, and the cotyledons of the embryo become filled with this stored food.

At maturity, seeds of these species have no endosperm and are termed exalbuminous seeds. Some exalbuminous seeds are bean, pea, oak, walnut, squash, sunflower, and radish. Seeds with an endosperm at maturity are termed albuminous seeds. Most monocots (e.g. grasses and palms) and many dicots (e.g. Brazil nut and castor bean) have albuminous seeds. All gymnosperm seeds are albuminous.

The seed coat (the testa) develops from the tissue, the integument, originally surrounding the ovule. The seed coat in the mature seed can be a paper-thin layer (e.g. peanut) or something more substantial (e.g. thick and hard in honey locust and coconut). The seed coat helps protect the embryo from mechanical injury and from drying out.

In addition to the three basic seed parts, some seeds have an appendage on the seed coat such an aril (as in yew and nutmeg) or an elaiosome (as in *Corydalis*) or hairs (as in cotton). A scar also may remain on the seed coat, called the hilum, where the seed was attached to the ovary wall by the funiculus.

Kinds of Seeds

Many structures commonly referred to as "seeds" are actually dry fruits. Sunflower seeds are sold commercially while still enclosed within the hard wall of the fruit, which must be split open to reach the seed. Different groups of plants have other modifications, the so-called stone fruits (such as the peach) have a hardened fruit layer (the endocarp) fused to and surrounding the actual seed. Nuts are the one-seeded, hard shelled fruit, of some plants, with an indehiscent seed, such as an acorn or hazelnut.

Seed Production

Seeds are produced in several related groups of plants, and their manner of production distinguishes the angiosperms ("enclosed seeds") from the gymnosperms ("naked seeds"). Angiosperm seeds are produced in a hard or fleshy structure called a fruit that encloses the seeds, hence the name. (Some fruits have layers of both hard and fleshy material). In gymnosperms, no special structure develops to enclose the seeds, which begin their development "naked" on the bracts of cones. However, the seeds do become covered by the cone scales as they develop in some species of conifer.

Seed production in natural plant populations vary widely from year-to-year in response to weather variables, insects and diseases, and internal cycles within the plants themselves. Over a 20-year period, for example, forests composed of loblolly pine and shortleaf pine produced from 0 to nearly 5 million sound pine seeds per hectare. Over this period, there were six bumper, five poor, and nine good seed crops, when evaluated in regard to producing adequate seedlings for natural forest reproduction.

Seed Development

The seed, which is an embryo with two points of growth (one of which forms the stem, the other the roots) is enclosed in a seed coat with some food reserves. Angiosperm seeds consist of three genetically distinct constituents: (1) the embryo formed from the zygote, (2) the endosperm, which is normally triploid, (3) the seed coat from tissue derived from the maternal tissue of the ovule. In angiosperms, the process of seed development begins with double fertilization and involves the fusion of the egg and sperm nuclei into a zygote.

The second part of this process is the fusion of the polar nuclei with a second sperm cell nucleus, thus forming a primary endosperm. Right after fertilization, the zygote is mostly inactive, but the primary endosperm divides rapidly to form the endosperm tissue. This tissue becomes the food the young plant will consume until the roots have developed after germination.

The seed coat forms from the two integuments or outer layers of cells of the ovule, which derive from tissue from the mother plant, the inner integument forms the tegmen and the outer forms the testa. When the seed coat forms from only one layer, it is also called the testa, though not all such testae are homologous from one species to the next.

In gymnosperms, the two sperm cells transferred from the pollen do not develop seed by double fertilization, but one sperm nucleus unites with the egg nucleus and the other sperm is not used. Sometimes each sperm fertilizes an egg cell and one zygote is then aborted or absorbed during early development. The seed is composed of the embryo (the result of fertilization) and tissue from the mother plant, which also form a cone around the seed in coniferous plants such as pine and spruce.

The ovules after fertilization develop into the seeds; the main parts of the ovule are the funicle; which attaches the ovule to the placenta, the nucellus; the main region of the ovule where the megagametophyte develops, the micropyle; a small pore or opening in the ovule where the pollen tube usually enters during the process of fertilization, and the chalaza; the base of the ovule opposite the micropyle, where integument and nucellus are joined together. The shape of the ovules as they develop often affects the final shape of the seeds. Plants generally produce ovules of four shapes: the most common shape is called anatropous, with a curved shape. Orthotropous ovules are straight with all the parts of the ovule lined up in a long row producing an uncurved seed. Campylotropous ovules have a curved megagametophyte often giving the seed a tight "C" shape. The last ovule shape is called amphitropous, where the ovule is partly inverted and turned back 90 degrees on its stalk (the funiculus).

In the majority of flowering plants, the zygote's first division is transversely oriented in regards to the long axis, and this establishes the polarity of the embryo. The upper or chalazal pole becomes the main area of growth of the embryo, while the lower or micropylar pole produces the stalk-like suspensor that attaches to the micropyle. The suspensor absorbs and manufacturers nutrients from the endosperm that are used during the embryo's growth.

The embryo is composed of different parts; the epicotyle will grow into the shoot, the radicle grows into the primary root, the hypocotyl connects the epicotyle and the radicle, the cotyledons form the seed leaves. Monocotyledonous plants have other structures; instead of the hypocotyle-epicotyle, it has a coleoptile that forms the first leaf and connects to the coleorhiza that connects to the primary root and adventitious roots form from the sides.

The seeds of corn are constructed with these structures; pericarp, scutellum (single large cotyledon) that absorbs nutrients from the endosperm, endosperm, plumule, radicle, coleoptile and coleorhiza - these last two structures are sheath-like and enclose the plumule and radicle, acting as a protective covering. The testae or seed coats of both monocots and dicots are often marked with patterns and textured markings, or have wings or tufts of hair.

Seed Size and Seed Set

Seeds are very diverse in size. The dust-like orchid seeds are the smallest, with about one million seeds per gram; they are often embryonic seeds with immature embryos and no significant energy reserves. Orchids and a few other groups of plants are mycoheterotrophs which depend on mycorrhizal fungi for nutrition during germination and the early growth of the seedling. Some terrestrial orchid seedlings, in fact, spend the first few years of their lives deriving energy from the fungi and do not produce green leaves. At over 20 kg, the largest seed is the *coco de mer*. Plants that produce smaller seeds can generate many more seeds per flower, while plants with larger seeds invest more resources into those seeds and normally produce fewer seeds. Small seeds are quicker to ripen and can be dispersed sooner, so fall blooming plants often have small seeds. Many annual plants produce great quantities of smaller seeds; this helps to ensure at least a few will end in a favourable place for growth. Herbaceous perennials and woody plants often have larger seeds; they can produce seeds over many years, and larger seeds have more energy reserves for germination and seedling growth and produce larger, more established seedlings after germination.

Seed Functions

Seeds serve several functions for the plants that produce them. Key among these functions are nourishment of the embryo, dispersal to a new location, and dormancy during unfavourable conditions. Seeds fundamentally are a means of reproduction and most seeds are the product of sexual reproduction which produces a remixing of genetic material and phenotype variability that natural selection acts on.

Embryo Nourishment

Seeds protect and nourish the embryo or young plant. Seeds usually give a seedling a faster start than a sporeling from a spore, because of the larger food reserves in the seed and the multicellularity of the enclosed embryo.

Seed Dispersal

Unlike animals, plants are limited in their ability to seek out favourable conditions for life and growth. As a result, plants have evolved many ways to disperse their offspring by dispersing their seeds. A seed must somehow “arrive” at a location and be there at a time favourable for germination and growth. When the fruits open and release their seeds in a regular way, it is called dehiscent, which is often distinctive for related groups of plants; these fruits include capsules, follicles, legumes, silicles and siliques. When fruits do not open and release their seeds in a regular fashion they are called indehiscent, which include the fruits achenes, caryopsis, nuts, samaras, and utricles. Seed dispersal is seen most obviously in fruits; however many seeds aid in their own dispersal. Some kinds of seeds are dispersed while still inside a fruit or cone, which later opens or disintegrates to release the seeds. Other seeds are expelled or released from the fruit prior to dispersal. For example, milkweeds produce a fruit type, known as a *follicle*, that splits open along one side to release the seeds. Iris capsules split into three "valves" to release their seeds.

By Wind (Anemochory)

Figure: *The seed pod of milkweed (*Asclepias syriaca*)*

- Some seeds (e.g., pine) have a wing that aids in wind dispersal.
- The dustlike seeds of orchids are carried efficiently by the wind.
- Some seeds, (e.g. milkweed, poplar) have hairs that aid in wind dispersal.

Other seeds are enclosed in fruit structures that aid wind dispersal in similar ways:

- Dandelion achenes have hairs.
- Maple samaras have two wings.

By Water (Hydrochory)

- Some plants, such as *Mucuna* and *Dioclea*, produce buoyant seeds termed sea-beans or drift seeds because they float in rivers to the oceans and wash up on beaches.

By Animals (Zoochory)

- Seeds (burrs) with barbs or hooks (e.g. acaena, burdock, dock) which attach to animal fur or feathers, and then drop off later.
- Seeds with a fleshy covering (e.g. apple, cherry, juniper) are eaten by animals (birds, mammals, reptiles, fish) which then disperse these seeds in their droppings.

- Seeds (nuts) which are an attractive long-term storable food resource for animals (e.g. acorns, hazelnut, walnut); the seeds are stored some distance from the parent plant, and some escape being eaten if the animal forgets them.

Myrmecochory is the dispersal of seeds by ants. Foraging ants disperse seeds which have appendages called elaiosomes (e.g. bloodroot, trilliums, Acacias, and many species of Proteaceae). Elaiosomes are soft, fleshy structures that contain nutrients for animals that eat them. The ants carry such seeds back to their nest, where the elaiosomes are eaten.

The remainder of the seed, which is hard and inedible to the ants, then germinates either within the nest or at a removal site where the seed has been discarded by the ants. This dispersal relationship is an example of mutualism, since the plants depend upon the ants to disperse seeds, while the ants depend upon the plants seeds for food. As a result, a drop in numbers of one partner can reduce success of the other. In South Africa, the Argentine ant (*Linepithema humile*) has invaded and displaced native species of ants. Unlike the native ant species, Argentine ants do not collect the seeds of *Mimetes cucullatus* or eat the elaiosomes. In areas where these ants have invaded, the numbers of *Mimetes* seedlings have dropped.

Seed Dormancy

Seed dormancy has two main functions: the first is synchronizing germination with the optimal conditions for survival of the resulting seedling; the second is spreading germination of a batch of seeds over time so that a catastrophe after germination (e.g. late frosts, drought, herbivory) does not result in the death of all offspring of a plant (bet-hedging). Seed dormancy is defined as a seed failing to germinate under environmental conditions optimal for germination, normally when the environment is at a suitable temperature with proper soil moisture.

This true dormancy or innate dormancy is therefore caused by conditions within the seed that prevent germination. Thus dormancy is a state of the seed, not of the environment. Induced dormancy, enforced dormancy or seed quiescence occurs when a seed fails to germinate because the external environmental conditions are inappropriate for germination, mostly in response to conditions being too dark or light, too cold or hot, or too dry.

Seed dormancy is not the same as seed persistence in the soil or on the plant, though even in scientific publications dormancy and persistence are often confused or used as synonyms.

Often seed dormancy is divided into four major categories: exogenous; endogenous; combinational; and secondary. A more recent system distinguishes five classes of dormancy: morphological, physiological, morphophysiological, physical and combinational dormancy.

Exogenous dormancy is caused by conditions outside the embryo including:

- Physical dormancy or hard seed coats occurs when seeds are impermeable to water. At dormancy break a specialized structure, the 'water gap', is disrupted in response to environmental cues, especially temperature, so that water can enter the seed and germination can occur. Plant families where physical dormancy occurs include Anacardiaceae, Cannaceae, Convulvulaceae, Fabaceae and Malvaceae.
- Chemical dormancy considers species that lack physiological dormancy, but where a chemical prevents germination. This chemical can be leached out of the seed by rainwater or snow melt or be deactivated somehow. Leaching of chemical inhibitors from the seed by rain water is often cited as an important cause of dormancy release in seeds of desert plants, however little evidence exists to support this claim.

Endogenous dormancy is caused by conditions within the embryo itself, including:

- Morphological dormancy where germination is prevented due to morphological characteristics of the embryo. In some species the embryo is just a mass of cells when seeds are dispersed, it is not differentiated. Before germination can take place both differentiation and growth of the embryo have to occur. In other species the embryo is differentiated but not fully grown (underdeveloped) at dispersal and embryo growth up to a species specific length is required before germination can occur. Examples of plant families where morphological dormancy occurs are Apiaceae, Cycadaceae, Liliaceae, Magnoliaceae and Ranunculaceae.
- Morphophysiological dormancy seeds with underdeveloped embryos, and which in addition have physiological components to dormancy. These seeds therefore require a dormancy-breaking treatments as well as a period of time to develop fully grown embryos. Plant families where morphophysiological dormancy occurs include Apiaceae, Aquifoliaceae, Liliaceae,

Magnoliaceae, Papaveraceae and Ranunculaceae. Some plants with morphophysiological dormancy like Asarum or Trillium species have multiple types of dormancy, one affects radicle (root) growth while the other affects plumule (shoot) growth. The terms "double dormancy" and "2-year seeds" are used for species whose seeds need two years to complete germination or at least two winters and one summer. Dormancy of the radicle (seedling root) is broken during the first winter after dispersal while dormancy of the shoot bud is broken during the second winter.

- Physiological dormancy means that the embryo can, due to physiological causes, not generate enough power to break through the seed coat, endosperm or other covering structures. Dormancy is typically broken at cool wet, warm wet or warm dry conditions. Abscisic acid is usually the growth inhibitor in seeds and its production can be affected by light.
    - o Drying; some plants including a number of grasses and those from seasonally arid regions need a period of drying before they will germinate. The seeds are released but need to have a lower moisture content before germination can begin. If the seeds remain moist after dispersal, germination can be delayed for many months or even years. Many herbaceous plants from temperate climate zones have physiological dormancy that disappears with drying of the seeds. Other species will germinate after dispersal only under very narrow temperature ranges, but as the seeds dry they are able to germinate over a wider temperature range.
- Combinational dormancy In seeds with combinational dormancy the seed or fruit coat is impermeable to water and the embryo has physiological dormancy. Depending on the species physical dormancy can be broken before or after physiological dormancy is broken.
- Secondary dormancy* is caused by conditions after the seed has been dispersed and occurs in some seeds when non-dormant seed is exposed to conditions that are not favourable to germination, very often high temperatures. The mechanisms of secondary dormancy are not yet fully understood but might involve the loss of sensitivity in receptors in the plasma membrane.

The following types of seed dormancy do not involve seed dormancy strictly spoken as lack of germination is prevented by the environment not by characteristics of the seed itself:

- Photodormancy or light sensitivity affects germination of some seeds. These photoblastic seeds need a period of darkness or light to germinate. In species with thin seed coats, light may be able to penetrate into the dormant embryo. The presence of light or the absence of light may trigger the germination process, inhibiting germination in some seeds buried too deeply or in others not buried in the soil.
- Thermodormancy is seed sensitivity to heat or cold. Some seeds, including cocklebur and amaranth, germinate only at high temperatures (30 °C or 86 °F); many plants that have seed that germinate in early to mid-summer have thermodormancy and germinate only when the soil temperature is warm. Other seeds need cool soils to germinate, while others, like celery, are inhibited when soil temperatures are too warm. Often thermodormancy requirements disappear as the seed ages or dries.

Not all seeds undergo a period of dormancy. Seeds of some mangroves are viviparous; they begin to germinate while still attached to the parent. The large, heavy root allows the seed to penetrate into the ground when it falls. Many garden plants have seeds that will germinate readily as soon as they have water and are warm enough; though their wild ancestors may have had dormancy, these cultivated plants lack seed dormancy. After many generations of selective pressure by plant breeders and gardeners, dormancy has been selected out.

For annuals, seeds are a way for the species to survive dry or cold seasons. Ephemeral plants are usually annuals that can go from seed to seed in as few as six weeks.

Seed Persistence and Seed Banks

Seed Germination

Seed germination is a process by which a seed embryo develops into a seedling. It involves the reactivation of the metabolic pathways that lead to growth and the emergence of the radicle or seed root and plumule or shoot. The emergence of the seedling above the soil surface is the next phase of the plant's growth and is called seedling establishment.

Three fundamental conditions must exist before germination can occur. (1) The embryo must be alive, called seed viability. (2) Any dormancy requirements that prevent germination must be overcome. (3) The proper environmental conditions must exist for germination.

Figure: *Germinating sunflower seedlings.*

Seed viability is the ability of the embryo to germinate and is affected by a number of different conditions. Some plants do not produce seeds that have functional complete embryos, or the seed may have no embryo at all, often called empty seeds. Predators and pathogens can damage or kill the seed while it is still in the fruit or after it is dispersed.

Environmental conditions like flooding or heat can kill the seed before or during germination. The age of the seed affects its health and germination ability: since the seed has a living embryo, over time cells die and cannot be replaced. Some seeds can live for a long time before germination, while others can only survive for a short period after dispersal before they die.

Seed vigor is a measure of the quality of seed, and involves the viability of the seed, the germination percentage, germination rate and the strength of the seedlings produced.

The germination percentage is simply the proportion of seeds that germinate from all seeds subject to the right conditions for growth. The germination rate is the length of time it takes for the seeds to germinate. Germination percentages and rates are affected by seed viability, dormancy and environmental effects that impact on the seed and seedling. In agriculture and horticulture quality seeds have high viability, measured by germination percentage plus the rate of germination.

This is given as a percent of germination over a certain amount of time, 90% germination in 20 days, for example. 'Dormancy' is covered above; many plants produce seeds with varying degrees of dormancy, and different seeds from the same fruit can have different degrees of dormancy. It's possible to have seeds with no dormancy if they are dispersed right away and do not dry (if the seeds dry they go into physiological dormancy). There is great variation amongst plants and a dormant seed is still a viable seed even though the germination rate might be very low.

Environmental conditions effecting seed germination include; water, oxygen, temperature and light.

Three distinct phases of seed germination occur: water imbibition; lag phase; and radicle emergence.

In order for the seed coat to split, the embryo must imbibe (soak up water), which causes it to swell, splitting the seed coat. However, the nature of the seed coat determines how rapidly water can penetrate and subsequently initiate germination.

The rate of imbibition is dependent on the permeability of the seed coat, amount of water in the environment and the area of contact the seed has to the source of water. For some seeds, imbibing too much water too quickly can kill the seed. For some seeds, once water is imbibed the germination process cannot be stopped, and drying then becomes fatal. Other seeds can imbibe and lose water a few times without causing ill effects, but drying can cause secondary dormancy.

Inducing Germination

A number of different strategies are used by gardeners and horticulturists to break seed dormancy. Scarification which allows water and gases to penetrate into the seed, include methods that physically break the hard seed coats or soften them by chemicals. Means of scarification include soaking in hot water or poking holes in the seed with a pin or rubbing them on sandpaper or cracking with a press or hammer. Soaking the seeds in solvents or acids is also effective for many seeds.

Sometimes fruits are harvested while the seeds are still immature and the seed coat is not fully developed and sown right away before the seed coat become impermeable. Under natural conditions seed coats are worn down by rodents chewing on the seed, the seeds rubbing against rocks (seeds are moved by the wind or water currents), by undergoing freezing and thawing of surface water, or passing through an animal's digestive tract.

In the latter case, the seed coat protects the seed from digestion, while often weakening the seed coat such that the embryo is ready to sprout when it gets deposited (along with a bit of fertilizer) far from the parent plant. Microorganisms are often effective in breaking down hard seed coats and are sometimes used by people as a treatment, the seeds are stored in a moist warm sandy medium for several months under non-sterile conditions.

Stratification also called moist-chilling is a method to break down physiological dormancy and involves the addition of moisture to the seeds so they imbibe water and then the seeds are subject to a period of moist chilling to after-ripen the embryo. Sowing outside in late summer and fall and allowing to overwinter outside under cool conditions is an effective way to stratify seeds, some seeds respond more favourably to periods of oscillating temperatures which are part of the natural environment.

Leaching or the soaking in water removes chemical inhibitors in some seeds that prevent germination. Rain and melting snow naturally accomplish this task. For seeds planted in gardens, running water is best - if soaked in a container, 12 to 24 hours of soaking is sufficient. Soaking longer, especially in stagnant water that is not changed, can result in oxygen starvation and seed death. Seeds with hard seed coats can be soaked in hot water to break open the impermeable cell layers that prevent water intake.

Other methods used to assist in the germination of seeds that have dormancy include prechilling, predrying, daily alternation of temperature, light exposure, potassium nitrate, the use of plant growth regulators like gibberellins, cytokinins, ethylene, thiourea, sodium hypochlorite plus others. Some seeds germinate best after a fire, for some seeds fire cracks hard seed coats while in other seeds chemical dormancy is broken in reaction to the presence of smoke, liquid smoke is often used by gardeners to assist in the germination of these species.

Origin and Evolution

The origin of seed plants is a problem that still remains unsolved. However, more and more data tends to place this origin in the middle Devonian. The description in 2004 of the proto-seed *Runcaria heinzelinii* in the Givetian of Belgium is an indication of that ancient origin of seed-plants. As with modern ferns, most land plants before this time reproduced by sending spores into the air, that would land and become whole new plants.

The first "true" seeds are described from the upper Devonian, which is probably the theater of their true first evolutionary radiation. The seed plants progressively became one of the major elements of nearly all ecosystems.

Economic Importance

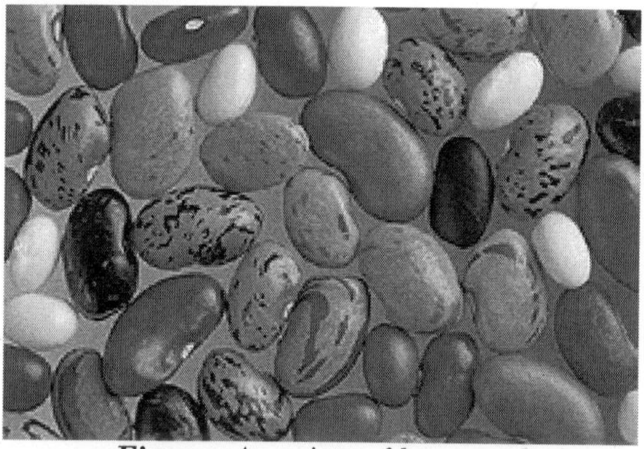

***Figure:** A variety of bean seeds.*

Edible seeds

Many seeds are edible and the majority of human calories comes from seeds, especially from cereals, legumes and nuts. Seeds also provide most cooking oils, many beverages and spices and some important food additives. In different seeds the seed embryo or the endosperm dominates and provides most of the nutrients. The storage proteins of the embryo and endosperm differ in their amino acid content and physical properties. For example the gluten of wheat, important in providing the elastic property to bread dough is strictly an endosperm protein.

Seeds are used to propagate many crops such as cereals, legumes, forest trees, turfgrasses and pasture grasses. Particularly in developing countries, a major constraint faced is the inadequacy of the marketing channels to get the seed to poor farmers. Thus the use of farmer-retained seed remains quite common. Seeds are also eaten by animals, and are fed to livestock. Many seeds are used as birdseed.

Poison and Food Safety

While some seeds are edible, others are harmful, poisonous or deadly. Plants and seeds often contain chemical compounds to discourage herbivores and seed predators. In some cases, these compounds simply taste bad (such as in mustard), but other compounds are toxic or break down into toxic compounds within the digestive system. Children, being smaller than adults, are more susceptible to poisoning by plants and seeds.

A deadly poison, ricin, comes from seeds of the castor bean. Reported lethal doses are anywhere from two to eight seeds, though only a few deaths have been reported when castor beans have been ingested by animals. In addition, seeds containing amygdalin—apple, apricot, bitter almond, peach, plum, cherry, quince, and others—when consumed in sufficient amounts, may cause Cyanide poisoning. Other seeds that contain poisons include annona, cotton, custard apple, datura, uncooked durian, golden chain, horse-chestnut, larkspur, locoweed, lychee, nectarine, rambutan, rosary pea, sour sop, sugar apple, wisteria, and yew. The seeds of the strychnine tree are also poisonous, containing the poison strychnine.

The seeds of many legumes, including the common bean (*Phaseolus vulgaris*), contain proteins called lectins which can cause gastric distress if the beans are eaten without cooking. The common bean and many others, including the soybean, also contain trypsin inhibitors which interfere with the action of the digestive enzyme trypsin. Normal cooking processes degrade lectins and trypsin inhibitors to harmless forms. Please see the category plant toxins for further relevant articles.

Other Uses

Cotton fibre grows attached to cotton plant seeds. Other seed fibres are from kapok and milkweed.

Many important nonfood oils are extracted from seeds. Linseed oil is used in paints. Oil from jojoba and crambe are similar to whale oil.

Seeds are the source of some medicines including castor oil, tea tree oil and the discredited cancer drug, Laetrile.

Figure: *Flax seed oil (in bottles) and coconut oil (in jars in the middle).*

Many seeds have been used as beads in necklaces and rosaries including Job's tears, Chinaberry, rosary pea, and castor bean. However, the latter three are also poisonous.

Other seed uses include:

- Seeds once used as weights for balances.
- Seeds used as toys by children, such as for the game Conkers.
- Resin from *Clusia rosea* seeds used to caulk boats.
- Nematicide from milkweed seeds.
- Cottonseed meal used as animal feed and fertilizer.

Seed Records

- The oldest viable carbon-14-dated seed that has grown into a plant was a Judean date palm seed about 2,000 years old, recovered from excavations at Herod the Great's palace on Masada in Israel. It was germinated in 2005. (A reported regeneration of *Silene stenophylla* (narrow-leafed campion) from material preserved for 31,800 years in the Siberian permafrost was achieved using fruit tissue, not seed.)
- The largest seed is produced by the coco de mer, or "double coconut palm", *Lodoicea maldivica*. The entire fruit may weigh up to 23 kilograms (50 pounds) and usually contains a single seed.
- The earliest fossil seeds are around 365 million years old from the Late Devonian of West Virginia. The seeds are preserved immature ovules of the plant *Elkinsia polymorpha*.

Figure: *The massive fruit of the coco de mer.*

Seedling

A seedling is a young plant sporophyte developing out of a plant embryo from a seed. Seedling development starts with germination of the seed. A typical young seedling consists of three main parts: the radicle (embryonic root), the hypocotyl (embryonic shoot), and the cotyledons (seed leaves).

The two classes of flowering plants (angiosperms) are distinguished by their numbers of seed leaves: monocotyledons (monocots) have one blade-shaped cotyledon, whereas dicotyledons (dicots) possess two round cotyledons. Gymnosperms are more varied. For example, pine seedlings have up to eight cotyledons. The seedlings of some flowering plants have no cotyledons at all. These are said to be acotyledons.

Germination and Early Seedling Development

During germination, the young plant emerges from its protective seed coat with its radicle first, followed by the cotyledons. The radicle orients towards gravity, while the hypocotyl orients away from gravity and elongates through cell expansion to push the cotyledons out of the ground.

Photomorphogenesis and Etiolation

Dicot seedlings grown in the light develop short hypocotyls and open cotyledons exposing the epicotyl. This is also referred to as photomorphogenesis. In contrast, seedlings grown in the dark develop long hypocotyls and their cotyledons remain closed around the epicotyl in an *apical hook*. This is referred to as skotomorphogenesis or etiolation. Etiolated seedlings are yellowish in colour as chlorophyll synthesis and chloroplast development depend on light. They will open their cotyledons and turn green when treated with light.

In a natural situation, seedling development starts with skotomorphogenesis while the seedling is growing through the soil and attempting to reach the light as fast as possible. During this phase, the cotyledons are tightly closed and form the *apical hook* to protect the shoot apical meristem from damage while pushing through the soil. In many plants, the seed coat still covers the cotyledons for extra protection.

Upon breaking the surface and reaching the light, the seedling's developmental programme is switched to photomorphogenesis. The cotyledons open upon contact with light (splitting the seed coat open, if still present) and become green, forming the first photosynthetic organs of the young plant. Until this stage, the seedling lives off the energy reserves stored in the seed. The opening of the cotyledons exposes the shoot apical meristem and the *plumule* consisting of the first *true leaves* of the young plant.

The seedlings sense light through the light receptors phytochrome (red and far-red light) and cryptochrome (blue light). Mutations in these photo receptors and their signal transduction components lead to seedling development that is at odds with light conditions, for example seedlings that show photomorphogenesis when grown in the dark.

Seedling Growth and Maturation

Once the seedling starts to photosynthesize, it is no longer dependent on the seed's energy reserves. The apical meristems start growing and give rise to the root and shoot. The first "true" leaves expand and can often be distinguished from the round cotyledons through their species-dependent distinct shapes. While the plant is growing and developing additional leaves, the cotyledons eventually senesce and fall off. Seedling growth is also affected by mechanical stimulation, such as by wind or other forms of physical contact, through a process called thigmomorphogenesis.

Pests and Diseases

Seedlings are particularly vulnerable to attack by pests and diseases, and can consequently experience high mortality rates. Pests and diseases which are especially damaging to seedlings include damping off, cutworms, slugs and snails.

Spore

In biology, a spore is a reproductive structure, some of which are adapted for dispersal and surviving for extended periods of time in unfavourable conditions. The spores of seed plants, however, are produced internally, and it is larger structures derived from the spores that disperse (the megaspores are formed within the ovules, and the microspores later give rise to pollen grains).

Spores form part of the life cycles of many bacteria, plants, algae, fungi and some protozoa. According to scientist Dr. Steinn Sigurdsson, "There are viable bacterial spores that have been found that are 40 million years old on Earth - and we know they're very hardened to radiation."

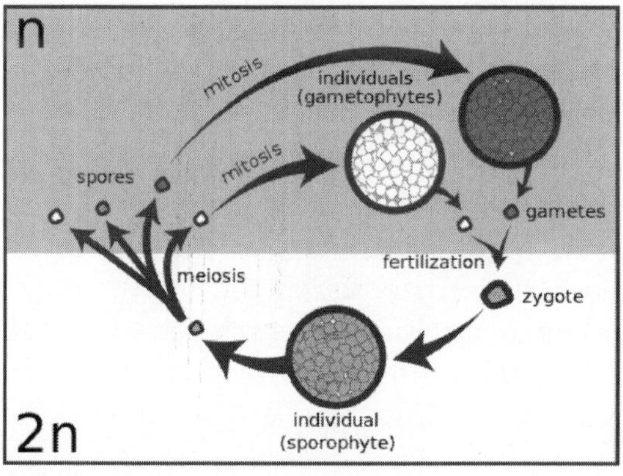

Figure: *Spores produced in a sporic life cycle.*

Spores are usually haploid and unicellular and are produced by meiosis in the sporangium of the sporophyte. Once conditions are favourable, the spore can develop into a new organism using mitotic division, producing a multicellular gametophyte, which eventually goes on to produce gametes. Two gametes fuse to create a new sporophyte. This cycle is known as alternation of generations.

Haploid spores produced by mitosis (known as mitospores) are used by many fungi for asexual reproduction. Spores are units of asexual reproduction; by contrast, gametes are units of sexual reproduction, as two gametes need to fuse to create a new organism.

Definition

The term *spore* derives from the ancient Greek word *spora*, meaning "seed, sowing," related to *sporos*, "sowing," and *speirein*, "to sow."

In common parlance, the difference between a "spore" and a "gamete" (both together called gonites) is that a spore will germinate and develop into a sporeling, while a gamete needs to combine with another gamete before developing further. However, the terms are somewhat interchangeable when referring to gametes.

A chief difference between spores and seeds as dispersal units is that spores have little food storage compared with seeds, and thus require more favourable conditions in order to successfully germinate. (This is not without its exceptions, however: many orchid seeds, although multicellular, are microscopic and lack endosperm, and spores of some fungi in the Glomeromycota commonly exceed 300 μm in diameter.) Seeds, therefore, are more resistant to harsh conditions and require less energy to start mitosis. Spores are produced in large numbers to increase the chance of a spore surviving in a number of notable examples.

Classification of Spore-producing Organisms

Vascular plant spores are always haploid. Vascular plants are either homosporous (or isosporous) or heterosporous. Plants that are homosporous produce spores of the same size and type. Heterosporous plants, such as seed plants, spikemosses, quillworts, and some aquatic ferns produce spores of two different sizes: the larger spore (megaspore) in effect functioning as a "female" spore and the smaller (microspore) functioning as a "male".

Classification of Spores

Spores can be classified in several ways:

By Spore-producing Structure

In fungi and fungus-like organisms, spores are often classified by the structure in which meiosis and spore production occurs. Since fungi are often classified according to their spore-producing structures, these spores are often characteristic of a particular taxon of the fungi.

- *Sporangiospores:* spores produced by a sporangium in many fungi such as zygomycetes.
- *Zygospores:* spores produced by a zygosporangium, characteristic of zygomycetes.
- *Ascospores:* spores produced by an ascus, characteristic of ascomycetes.
- *Basidiospores:* spores produced by a basidium, characteristic of basidiomycetes.
- *Aeciospores:* spores produced by a aecium in some fungi such as rusts or smuts.
- *Urediospores:* spores produced by a uredinium in some fungi such as rusts or smuts.
- *Teliospores:* spores produced by a telium in some fungi such as rusts or smuts.
- *Oospores:* spores produced by a oogonium, characteristic of oomycetes.
- *Carpospores:* spores produced by a carposporophyte, characteristic of red algae.
- *Tetraspores:* spores produced by a tetrasporophyte, characteristic of red algae.

By Function

- *Chlamydospores:* thick-walled resting spores of fungi produced to survive unfavourable conditions.
- Parasitic fungal spores may be classified into internal spores, which germinate within the host, and external spores, also called environmental spores, released by the host to infest other hosts.

By Origin During Life Cycle

- Meiospores: spores produced by meiosis; they are thus haploid, and give rise to a haploid daughter cell(s) or a haploid individual. Examples are the precursor cells of gametophytes of seed plants found in flowers (angiosperms) or cones (gymnosperms).
    - o Microspores: meiospores that give rise to a male gametophyte, (pollen in seed plants).

  - o Megaspores (or macrospores): meiospores that give rise to a female gametophyte, (in seed plants the gametophyte forms within the ovule).
- Mitospores (or conidia, conidiospores): spores produced by mitosis; they are characteristic of Ascomycetes. Fungi in which only mitospores are found are called "mitosporic fungi" or "anamorphic fungi", and are previously classified under the taxon Deuteromycota.

By Motility

Spores can be differentiated by whether they can move or not.

- *Zoospores:* mobile spores that move by means of one or more flagella, and can be found in some algae and fungi.
- *Aplanospores:* immobile spores that may nevertheless potentially grow flagella.
- *Autospores:* immobile spores that cannot develop flagella.
- *Ballistospores:* spores that are actively discharged from the body of the fungal fruiting body. Most basidiospores are also ballistospores, and another notable example is spores of *Pilobolus*.
- *Statismospores:* spores that are not actively discharged from the fungal fruiting body. Examples are puffballs.

Anatomy

Under high magnification, spores can be categorized as either monolete spores or trilete spores. In monolete spores, there is a single line on the spore indicating the axis on which the mother spore was split into four along a vertical axis. In trilete spores, all four spores share a common origin and are in contact with each other, so when they separate, each spore shows three lines radiating from a centre pole.

Trilete Spores

Trilete spores, formed by the dissociation of a spore tetrad, are taken as the earliest evidence of life on land, dating to the mid-Ordovician (early Llanvirn, ~470 million years ago).

Dispersal

In fungi, both asexual and sexual spores or sporangiospores of many fungal species are actively dispersed by forcible ejection from their reproductive structures. This ejection ensures exit of the spores from the reproductive structures as well as travelling through the air over long distances. Many fungi thereby possess specialized mechanical and physiological mechanisms as well as spore-surface structures, such as hydrophobins, for spore ejection. These mechanisms include, for example, forcible discharge of ascospores enabled by the structure of the ascus and accumulation of osmolytes in the fluids of the ascus that lead to explosive discharge of the ascospores into the air. The forcible discharge of single spores termed *ballistospores* involves formation of a small drop of water (Buller's drop), which upon contact with the spore leads to its projectile release with an initial acceleration of more than 10,000 g. Other fungi rely on alternative mechanisms for spore release, such as external mechanical forces, exemplified by puffballs. Attracting insects, such as flies, to fruiting structures, by virtue of their having lively colours and a putrid odour, for dispersal of fungal spores is yet another strategy, most prominently used by the stinkhorns.

In the case of spore-shedding vascular plants such as ferns, wind distribution of very light spores provides great capacity for dispersal. Also, spores are less subject to animal predation than seeds because they contain almost no food reserve; however they are more subject to fungal and bacterial predation. Their chief advantage is that, of all forms of progeny, spores require the least energy and materials to produce. In the spikemoss *Selaginella lepidophylla*, dispersal is achieved in part by an unusual type of diaspore, a tumbleweed.

Tree

A tree is a perennial woody plant. It most often has many secondary branches supported clear of the ground on a single main stem or trunk with clear apical dominance. A minimum height specification at maturity is cited by some authors, varying from 3 m to 6 m; some authors set a minimum of 10 cm trunk diameter (30 cm girth). Woody plants that do not meet these definitions by having multiple stems and/or small size are usually called shrubs, although many trees such as Mallee do not meet such definitions. Compared with most other plants, trees are long-lived, some reaching several thousand years old and growing to up to 115 m (379 ft) high.

Trees are an important component of the natural landscape because of their prevention of erosion and the provision of a weather-sheltered ecosystem in and under their foliage. They also play an important role in producing oxygen and reducing carbon dioxide in the atmosphere, as well as moderating ground temperatures. They are also elements in landscaping and agriculture, both for their aesthetic appeal and their orchard crops (such as apples). Wood from trees is a building material, as well as a primary energy source in many developing countries. Trees also play a role in many of the world's mythologies.

Classification

A tree is a plant form that occurs in many different orders and families of plants. Most species of trees today are flowering plants (angiosperms) and conifers. Trees show a variety of growth forms, leaf type and shape, bark characteristics and reproductive organs. For the listing of examples of well-known trees and how they are classified. The tree form has evolved separately in unrelated classes of plants, in response to similar environmental challenges, making it a classic example of parallel evolution. With an estimate of 100,000 tree species, the number of tree species worldwide might total 25 percent of all living plant species.

The majority of tree species grow in tropical regions of the world and many of these areas have not been surveyed yet by botanists, making species diversity and ranges poorly understood. The earliest tree-like organisms were tree ferns, horsetails and lycophytes, which grew in forests in the Carboniferous period, however these plants were not trees, since they lacked woody tissue. Trees evolved in the Triassic period, with conifers, ginkgos, cycads and other gymnosperms appeared producing woody tissue, and were subsequently followed by tree-form flowering plants in the Cretaceous period.

A small group of trees growing together is called a grove or copse, and a landscape covered by a dense growth of trees is called a forest. Several biotopes are defined largely by the trees that inhabit them; examples are rainforest and taiga. A landscape of trees scattered or spaced across grassland (usually grazed or burned over periodically) is called a savanna. A forest of great age is called old growth forest or ancient woodland (in the UK). A young tree is called a sapling.

Morphology

The parts of a tree are the roots, trunk(s), branches, twigs and leaves. Tree stems consist mainly of support and transport tissues (xylem and phloem). Wood consists of *xylem* cells, and bark is made of *phloem* and other tissues external to the vascular cambium. Trees may be grouped into *exogenous* and *endogenous* trees according to the way in which their stem diameter increases. Exogenous trees, which comprise the great majority of trees (all conifers, and almost all broadleaf trees), grow by the addition of new wood outwards, immediately under the bark. Endogenous trees, mainly in the monocotyledons (e.g., aloes and dragon trees), grow by addition of new material as discrete bundles within the existing trunk tissue. Trees can be identified to genus or species by a combination of the tree's shape, and the characteristics of its bark, leaves, flowers, and fruit. The leaves may be either deciduous or evergreen.

As an exogenous tree grows, it creates growth rings as new wood is laid down concentrically over the old wood. In species growing in areas with seasonal climate changes, wood growth produced at different times of the year may be visible as alternating light and dark, or soft and hard, rings of wood. In temperate climates, and tropical climates with a single wet-dry season alternation, the growth rings are annual, each pair of light and dark rings being one year of growth; these are known as annual rings.

In areas with two wet and dry seasons each year, there may be two pairs of light and dark rings each year; and in some (mainly semi-desert regions with irregular rainfall), there may be a new growth ring with each rainfall. In tropical rainforest regions, with constant year-round climate, growth is continuous and the growth rings are not visible nor is there a change in the wood texture. In species with annual rings, these rings can be counted to determine the age of the tree, and used to date cores or even wood taken from trees in the past, a practice known as the science of dendrochronology. Very few tropical trees can be accurately dated in this manner. Age determination in this manner is also impossible in endogenous trees.

The roots of a tree are generally embedded in earth, providing anchorage for the above-ground biomass and absorbing water and nutrients from the soil. However, while ground nutrients are essential to a tree's growth the majority of its biomass – over 90 percent – comes from carbon dioxide absorbed from the atmosphere. Above ground, the trunk gives height to the leaf-bearing branches, aiding in competition with other plant species for sunlight. In many trees, the arrangement of the branches optimizes exposure of the leaves to sunlight.

Not all trees have all the plant organs or parts mentioned above. For example, the saguaro cactus of North America has no functional leaves. A plant form that is similar to a tree, but generally having smaller, multiple trunks and/or branches that arise near the ground, is called a shrub. However, no precise differentiation between shrubs and trees is possible. Given their small size, bonsai plants would not technically be "trees", but one should not confuse reference to the form of a species with the size or shape of individual specimens. A spruce seedling does not fit the definition of a tree, but all spruces are trees.

Record Breaking Trees

The world's champion trees can be rated on height, trunk diameter or girth, total size, and age.

Tallest Trees

The heights of the tallest trees in the world have been the subject of considerable dispute and much exaggeration. Modern verified measurements with laser rangefinders, or with tape drop measurements made by tree climbers (such as those carried out by canopy researchers or members of groups like the Native Tree Society), have shown that some older measuring methods and measurements are often unreliable, sometimes producing exaggerations of 5% to 15% or more above the real height. Historical claims of trees growing to 130 m (430 ft), and even 150 m (490 ft), are now largely disregarded as unreliable, and attributed to human error. Historical records of fallen trees measured prostrate on the ground are considered to be somewhat more reliable. The following are now accepted as the top nine tallest reliably measured species (taken only currently standing specimens):

1. Coast Redwood (*Sequoia sempervirens*): 115.66 m (379.46 ft), Hyperion, Redwood National Park, California, United States
2. Coast Douglas-fir (*Pseudotsuga menziesii*): 99.76 m (327.3 ft), Brummit Creek, Coos County, Oregon, United States
3. Australian Mountain Ash (*Eucalyptus regnans*): 99.6 m (327 ft), south of Hobart, Tasmania, Australia
4. Sitka Spruce (*Picea sitchensis*): 96.7 m (317 ft), Prairie Creek Redwoods State Park, California, United States
5. Giant Sequoia (*Sequoiadendron giganteum*): 95.8 m (314 ft), Sequoia National Forest, California, United States
6. Tasmanian Blue Gum (*Eucalyptus globulus*): 90.7 m (298 ft), Tasmania, Australia
7. Manna Gum (*Eucalyptus viminalis*): 89 m (292 ft), Evercreech Forest Reserve, Tasmania, Australia
8. *Shorea faguetiana*: 88.3 m (290 ft) Tawau Hills National Park, in Sabah on the island of Borneo
9. Alpine Ash (*Eucalyptus delegatensis*): 87.9 m (288 ft), Tasmania, Australia.

Stoutest Trees

The girth of a tree is usually much easier to measure than the height, as it is a simple matter of stretching a tape round the trunk, and pulling it taut to find the circumference. Despite this, UK tree author Alan Mitchell made the following comment about measurements of yew trees:

The aberrations of past measurements of yews are beyond belief. For example, the tree at Tisbury has a well-defined, clean, if irregular bole at least 1.5 m long. It has been found to have a girth that dilated and shrunk in the following way: 11.28 m (1834 Loudon), 9.3 m (1892 Lowe), 10.67 m (1903 Elwes and Henry), 9.0 m (1924 E. Swanton), 9.45 m (1959 Mitchell) ... Earlier measurements have therefore been omitted.

As a general standard, tree girth is taken at "breast height". This is cited as dbh (diameter at breast height) in tree and forestry literature. Breast height is defined differently in different situations, with most forestry measurements taking girth at 1.3 m above ground, while those who measure ornamental trees usually measure at 1.5 m above ground; in most cases this makes little difference to the measured girth. On sloping ground, the "above ground" reference point is usually taken as the highest point on the ground touching the trunk, but in North America a point, that is the average of the highest point and the lowest point the tree trunk appears to contact the soil, is usually used. Some of the inflated old measurements may have been taken at ground level. Some past exaggerated measurements also result from measuring the complete next-to-bark measurement, pushing the tape in and out over every crevice and buttress.

Modern trends are to cite the tree's diameter rather than the circumference. Diameter of the tree is calculated by finding the medium diameter of the trunk, in most cases obtained by dividing the measured circumference by ð; this assumes the trunk is mostly circular in cross-section (an oval or irregular cross-section would result in a mean diameter slightly greater than the assumed circle). Accurately measuring circumference or diameter is difficult in species with the large buttresses that are especially characteristic in many species of rainforest trees. Simple measurement of circumference of such trees can be misleading when the circumference includes much empty space between buttresses.

One further problem with measuring baobabs *Adansonia* is that these trees store large amounts of water in the very soft wood in their trunks. This leads to marked variation in their girth over the year (though not more than about 2.5%), swelling to a maximum at the end of the rainy season, minimum at the end of the dry season.

The stoutest living single-trunk species in diameter are:

1. Montezuma Cypress *Taxodium mucronatum*: 11.62 m (38.1 ft), Árbol del Tule, Santa Maria del Tule, Oaxaca, Mexico. Note though that this diameter includes buttressing; the actual idealised diameter of the area of its wood is 9.38 m (30.8 ft).
2. Giant Sequoia *Sequoiadendron giganteum*: 8.85 m (29.0 ft), General Grant tree, General Grant Grove, California, United States
3. Coast Redwood *Sequoia sempervirens*: 7.9 m (26 ft), Lost Monarch Jedediah Smith Redwoods State Park, California, United States.
4. Australian Oak *Eucalyptus obliqua*: 6.72 m (22.0 ft)
5. Australian Mountain Ash *Eucalyptus regnans*: 6.52 m (21.4 ft), Big Foot
6. Western Redcedar *Thuja plicata*: 5.99 m (19.7 ft), Kalaloch Cedar, Olympic National Park
7. Sitka Spruce *Picea sitchensis*: 5.39 m (17.7 ft), Quinalt Lake Spruce, Olympic National Park
8. Kauri *Agathis australis* (Te Matua Ngahere): 5.33 m (17.5 ft).
9. Alerce *Fitzroya cupressoides*: 5.0 m (16.4 ft).

An additional problem lies in instances where multiple trunks (whether from an individual tree or multiple trees) grow together. The Sacred Fig is a notable example of this, forming additional "trunks" by growing adventitious roots down from the branches, which then thicken up when the root reaches the ground to form new trunks; a single Sacred Fig tree can have hundreds of such trunks. The multi-stemmed Hundred Horse Chestnut was known to have a circumference of 57.9 m (190 ft) when it was measured in 1780.

Largest Trees

The largest trees in total volume are both tall and large in diameter and, in particular, hold a large diameter high up the trunk. Measurement is very complex, particularly if branch volume is to be included as well as the trunk volume, so measurements have only been made for a small number of trees, and generally only for the trunk. No attempt has ever been made to include root volume. Measuring standards vary.

The top ten species measured so far are*:

1. Giant Sequoia *Sequoiadendron giganteum*: 1,487 m^3 (52,508 cu ft), General Sherman
2. Coast Redwood *Sequoia sempervirens*: 1,203 m^3 (42,500 cu ft), Lost Monarch
3. Montezuma Cypress *Taxodium mucronatum*: 750 m^3 (25,000 cu ft), Árbol del Tule
4. Kauri *Agathis australis* (Tâne Mahuta): 516 m^3
5. Western Redcedar *Thuja plicata*: 500 m^3 (17,650 cu ft), Quinault Lake Redcedar
6. Tasmanian Blue Gum *Eucalyptus globulus*: 368 m^3 (13,000 cu ft), Rullah Longatyle (Strong Girl, also Grieving Giant)
7. Australian Mountain Ash *Eucalyptus regnans*: 360 m^3 (12,714 cu ft), Arve Big Tree
8. Coast Douglas-fir *Pseudotsuga menziesii* 349 m^3 (12,320 cu ft) Red Creek Tree
9. Sitka Spruce *Picea sitchensis* 337 m^3 (11,920 cu ft) Queets Spruce
10. Australian Oak *Eucalyptus obliqua*: 337 m^3 (11,920 cu ft) Gothmog
11. Alpine Ash *Eucalyptus delegatensis*: 286 m^3 (10,100 cu ft), located in Styx River Valley.

This list does not take into account now dead specimens.

Oldest Trees

The oldest trees are determined by growth rings, which can be seen if the tree is cut down, or in cores taken from the bark to the centre of the tree. Accurate determination is only possible for trees that produce growth rings, generally those in seasonal climates. Trees in uniform non-seasonal tropical climates grow continuously and do not have distinct growth rings. It is also only possible for trees that are solid to the centre. Many very old trees become hollow as the dead heartwood decays. For some of these species, age estimates have been made on the basis of extrapolating current growth rates, but the results are usually largely speculation. White (1998) proposes a method of estimating the age of large and veteran trees in the United Kingdom through the correlation between a tree's stem diameter, growth character and age.

The verified oldest measured ages are:

1. Great Basin Bristlecone Pine (Methuselah) *Pinus longaeva*: 4,844 years
2. Alerce *Fitzroya cupressoides*: 3,622 years
3. Giant Sequoia *Sequoiadendron giganteum*: 3,266 years
4. Sugi *Cryptomeria japonica*: 3,000 years
5. Huon-pine *Lagarostrobos franklinii*: 2,500 years.

Other species suspected of reaching exceptional age include European Yew *Taxus baccata* (probably over 2,000 years) and Western Redcedar *Thuja plicata*. The oldest known European Yew is the Llangernyw Yew in the Churchyard of Llangernyw village in North Wales, which is estimated to be between 4,000 and 5,000 years old.

Damage

The two sources of tree damage are either biotic (from living sources) or abiotic (from non-living sources). Biotic sources include insects that bore into the tree, deer that rub bark off, and fungi.

Abiotic sources include lightning, vehicles impacts, and construction activities. Construction activities can involve a number of damage sources, including grade changes that prevent aeration to roots, spills involving toxic chemicals such as cement or petroleum products, or severing of branches or roots.

Both damage sources can result in trees becoming dangerous, and the term "hazard trees" is commonly used by arborists, and industry groups such as power line operators. Hazard trees are trees that, due to disease or other factors, are more susceptible to falling in windstorms, or having parts of the tree fall.

Evaluating the danger a tree presents is based on a process called the Quantified Tree Risk Assessment. Assessment as to labelling a tree a hazard tree can be based on a field examination. Assessment as a result of construction activities that will damage a tree is based on three factors: severity, extent and duration. Severity relates usually to the degree of intrusion into the TPZ and resultant root loss. Extent is frequently a percentage of a factor such as canopy, roots or bark, and duration is normally based on time. Root severing is considered permanent in time.

Trees are similar to people. Both can withstand massive amounts of some types of damage and survive, but even small amounts of certain types of trauma can result in death. Arborists are very aware that established trees will not tolerate any appreciable disturbance of the root system. However, lay people and construction professionals seldom recognise how easily a tree can be killed.

One reason for confusion about tree damage from construction involves the dormancy of trees during winter. Another factor is that trees may not show symptoms of damage until two years or longer after damage has occurred. For that reason, persons uneducated in arboricultural science may not correlate the actual cause and resultant effect.

Various organisations, such as the International Society of Arboriculture, the British Standards Institute and the Tree Industry Association (formerly the National Arborist Association), have long recognised the importance of construction activities that impact tree health. The impacts are important because they can result in monetary losses due to tree damage and resultant remediation or replacement costs, as well as violation of government ordinances or community or subdivision restrictions.

As a result, protocols for tree management prior to, during and after construction activities are well established, tested and refined. These basic steps are involved:

- Review of the construction plans
- Development of the related tree inventory
- Application of standard construction tree management protocols
- Assessment of potential for expected tree damages
- Development of a tree protection plan (providing for pre-, concurrent, and post construction damage prevention and remediation steps)
- Development of a tree protection plan
- Development of a remediation plan
- Implementation of tree protection zones (TPZ)
- Assessment of construction tree damage, post-construction
- Implementation of the remediation plan.

Trees in Culture

The tree has always been a cultural symbol. Common icons are the World tree, for instance Yggdrasil, and the tree of life. The tree is often used to represent nature or the environment itself. The tree is also a symbol and mascot for Stanford University and Currier House (Harvard College).

Tree Value Approximation (USA)

Studies have shown that trees contribute as much as 27% of the appraised land value in certain markets and cite the following which can be extrapolated with care.

Table: *Basic Tree Values (Varies by Region)*

Diameter(inches)	*value(1985 US$)*
10	$1,729
14	$3,388
18	$5,588
26	$11,682
30	$15,554

These most likely use diameter measured at breast height, 4.5 feet (140 cm) above ground, not the larger base diameter. A general model for any year and diameter is Value = 17.27939 (year-1985) assuming 2.2% inflation per year. The right side of this equation is written to paste into Excel or Google to perform the calculation. Extrapolations from any model can vary widely, so value estimates for diameters larger than 30 inches (760 mm) may have to be capped so trees do not exceed 27% of total appraised land value.

Vine

A vine in the narrowest sense is the grapevine (*Vitis*), but more generally it can refer to any plant with a growth habit of trailing or scandent, that is to say climbing, stems or runners. The word also can refer to such stems or runners themselves, for instance when used in wicker work.

Climbing Plants

Certain plants always grow as vines, while a few grow as vines only part of the time. For instance, poison ivy and bittersweet can grow as low shrubs when support is not available, but will become vines when support is available.

A vine displays a growth form based on long stems. This has two purposes. A vine may use rock exposures, other plants, or other supports for growth rather than investing energy in a lot of supportive tissue, enabling the plant to reach sunlight with a minimum investment of energy. This has been a highly successful growth form for plants such as kudzu and Japanese honeysuckle, both of which are invasive exotics in parts of North America. There are some tropical vines that develop skototropism, and grow away from the light, a type of negative phototropism. Growth away from light allows the vine to reach a tree trunk, which it can then climb to brighter regions.

The vine growth form may also enable plants to colonize large areas quickly, even without climbing high. This is the case with periwinkle and ground ivy. It is also an adaptation to life in areas where small patches of fertile soil are adjacent to exposed areas with more sunlight but little or no soil. A vine can root in the soil but have most of its leaves in the brighter, exposed area, getting the best of both worlds.

A climbing habit has evolved independently in several plant families, using many different climbing methods. Some plants climb by twining their stems around a support (e.g., morning glories, *Ipomoea* species). Others climb by way of adventitious, clinging roots (e.g., ivy, *Hedera* species), with twining petioles (e.g., *Clematis* species), or using tendrils, which can be specialized shoots (Vitaceae), leaves (Bignoniaceae), or even influorescences (*Passiflora*). Others climb through the use of thorns, which pierce the support (e.g. climbing rose); or by other hooked structures, such as hooked branches (e.g. *Artabotrys hexapetalus*).

The climbing fetterbush (*Pieris phillyreifolia*) is a woody shrub-vine which climbs without clinging roots, tendrils, or thorns. It directs its stem into a crevice in the bark of fibrous barked trees (such as bald cypress) where the stem adopts a flattened profile and grows up the tree underneath the host tree's outer bark. The fetterbush then sends out branches that emerge near the top of the tree. Species of *Parthenocissus* (Vitaceae) produce twining tendrils that are modified stems, but which also produce adhesive pads at the end that attach themselves quite strongly to the support.

The evolution of a climbing habit has been implicated as a key innovation associated with the evolutionary success and diversification of a number of taxonomic groups of plants.Most vines are flowering plants. These may be divided into woody vines or lianas, such as wisteria, kiwifruit, and common ivy, and herbaceous (nonwoody) vines, such as morning glory.

One odd group of vining plants is the fern genus *Lygodium*, called climbing ferns. The stem does not climb, but rather the fronds (leaves) do. The fronds unroll from the tip, and theoretically never stop growing. In the meantime, they can form thickets as they unroll over other plants, rockfaces, and fences.

Use as Garden Plants

Gardeners can use the tendency of climbing plants to grow quickly. If a plant display is wanted quickly, a climber can achieve this. Climbers can be trained over walls, pergolas, fences, etc. Climbers can be grown over other plants to provide additional attraction. Artificial support can also be provided. Some climbers climb by themselves; others need work, such as tying them in and training them.

Wood

Wood is a hard, fibrous tissue found in many trees. It has been used for hundreds of thousands of years for both fuel and as a construction material. It is an organic material, a natural composite of cellulose fibres (which are strong in tension) embedded in a matrix of lignin which resists compression.

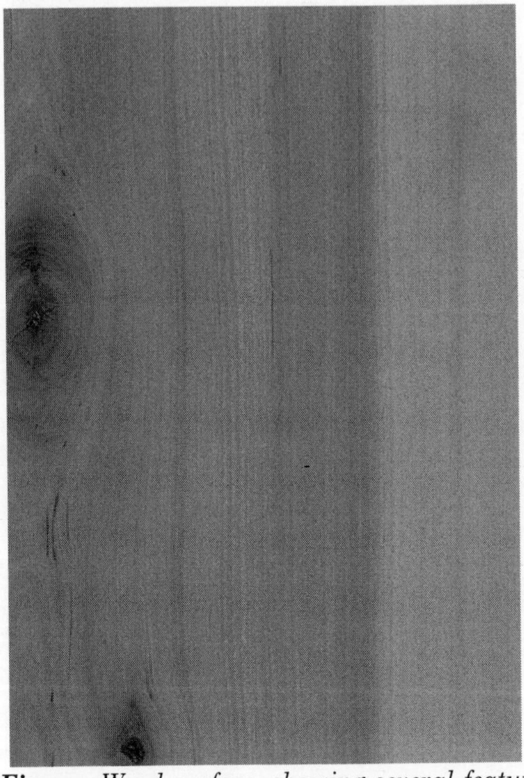

Figure: *Wood surface, showing several features*

Wood is sometimes defined as only the secondary xylem in the stems of trees, or it is defined more broadly to include the same type of tissue elsewhere such as in tree roots or in other plants such as shrubs. In a living tree it performs a support function, enabling woody plants to grow large or to stand up for themselves. It also mediates the transfer of water and nutrients to the leaves and other growing tissues. Wood may also refer to other plant materials with comparable properties, and to material engineered from wood, or wood chips or fibre.

The earth contains about one trillion tonnes of wood, which grows at a rate of 10 billion tonnes per year. As an abundant, carbon-neutral renewable resource, woody materials have been of intense interest as a source of renewable energy. In 1991, approximately 3.5 billion cubic meters of wood were harvested. Dominant uses were for furniture and building construction.

History

A 2011 discovery in the Canadian province of New Brunswick uncovered the earliest known plants to have grown wood, approximately 395 to 400 million years ago. People have used wood for millennia for many purposes, primarily as a fuel or as a construction material for making houses, tools, weapons, furniture, packaging, artworks, and paper.

Wood can be dated by carbon dating and in some species by dendrochronology to make inferences about when a wooden object was created. The year-to-year variation in tree-ring widths and isotopic abundances gives clues to the prevailing climate at that time.

Physical Properties

Growth Rings: Wood, in the strict sense, is yielded by trees, which increase in diameter by the formation, between the existing wood and the inner bark, of new woody layers which envelop the entire stem, living branches, and roots. This process is known as secondary growth; it is the result of cell division in the vascular cambium, a lateral meristem, and subsequent expansion of the new cells.

Where there are clear seasons, growth can occur in a discrete annual or seasonal pattern, leading to growth rings; these can usually be most clearly seen on the end of a log, but are also visible on the other surfaces. If these seasons are annual these growth rings are referred to as annual rings. Where there is no seasonal difference growth rings are likely to be indistinct or absent. If there are differences within a growth ring, then the part of a growth ring nearest the centre of the tree, and formed early in the growing season when growth is rapid, is usually composed of wider elements. It is usually lighter in colour than that near the outer portion of the ring, and is known as earlywood or springwood. The outer portion formed later in the season is then known as the latewood or summerwood. However, there are major differences, depending on the kind of wood.

Knots

A knot is a particular type of imperfection in a piece of wood; it will affect the technical properties of the wood, usually for the worse, but may be exploited for visual effect. In a longitudinally sawn plank, a knot will appear as a roughly circular "solid" (usually darker) piece of wood around which the grain of the rest of the wood "flows" (parts and rejoins). Within a knot, the direction of the wood (grain direction) is up to 90 degrees different from the grain direction of the regular wood.

In the tree a knot is either the base of a side branch or a dormant bud. A knot (when the base of a side branch) is conical in shape (hence the roughly circular cross-section) with the inner tip at the point in stem diameter at which the plant's vascular cambium was located when the branch formed as a bud.

During the development of a tree, the lower limbs often die, but may remain attached for a time, sometimes years. Subsequent layers of growth of the attaching stem are no longer intimately joined with the dead limb, but are grown around it. Hence, dead branches produce knots which are not attached, and likely to drop out after the tree has been sawn into boards.

In grading lumber and structural timber, knots are classified according to their form, size, soundness, and the firmness with which they are held in place. This firmness is affected by, among other factors, the length of time for which the branch was dead while the attaching stem continued to grow.

Knots materially affect cracking and warping, ease in working, and cleavability of timber. They are defects which weaken timber and lower its value for structural purposes where strength is an important consideration. The weakening effect is much more serious when timber is subjected to forces perpendicular to the grain and/or tension than where under load along the grain and/or compression.

The extent to which knots affect the strength of a beam depends upon their position, size, number, and condition. A knot on the upper side is compressed, while one on the lower side is subjected to tension. If there is a season check in the knot, as is often the case, it will offer little resistance to this tensile stress. Small knots, however, may be located along the neutral plane of a beam and increase the strength by preventing longitudinal shearing. Knots in a board or plank are least injurious when they extend through it at right angles to its broadest surface. Knots which occur near the ends of a beam do not weaken it. Sound knots which occur in the central portion one-fourth the height of the beam from either edge are not serious defects.

Knots do not necessarily influence the stiffness of structural timber, this will depend on the size and location. Stiffness and elastic strength are more dependent upon the sound wood than upon localized defects. The breaking strength is very susceptible to defects. Sound knots do not weaken wood when subject to compression parallel to the grain.

In some decorative applications, wood with knots may be desirable to add visual interest. In applications where wood is painted, such as skirting boards, fascia boards, door frames and furniture, resins present in the timber may continue to 'bleed' through to the surface of a knot for months or even years after manufacture and show as a yellow or brownish stain. A knot primer paint or solution, correctly applied during preparation, may do much to reduce this problem but it is difficult to control completely, especially when using mass-produced kiln-dried timber stocks.

Heartwood and Sapwood

Heartwood (or duramen) is wood that as a result of a naturally occurring chemical transformation has become more resistant to decay. Heartwood formation occurs spontaneously (it is a genetically programmed process). Once heartwood formation is complete, the heartwood is dead. Some uncertainty still exists as to whether heartwood is truly dead, as it can still chemically react to decay organisms, but only once.

Usually heartwood looks different; in that case it can be seen on a cross-section, usually following the growth rings in shape. Heartwood may (or may not) be much darker than living wood. It may (or may not) be sharply distinct from the sapwood. However, other processes, such as decay, can discolour wood, even in woody plants that do not form heartwood, with a similar colour difference, which may lead to confusion.

Sapwood (or alburnum) is the younger, outermost wood; in the growing tree it is living wood, and its principal functions are to conduct water from the roots to the leaves and to store up and give back according to the season the reserves prepared in the leaves. However, by the time they become competent to conduct water, all xylem tracheids and vessels have lost their cytoplasm and the cells are therefore functionally dead.

All wood in a tree is first formed as sapwood. The more leaves a tree bears and the more vigorous its growth, the larger the volume of sapwood required. Hence trees making rapid growth in the open have thicker sapwood for their size than trees of the same species growing in dense forests. Sometimes trees (of species that do form heartwood) grown in the open may become of considerable size, 30 cm or more in diameter, before any heartwood begins to form, for example, in second-growth hickory, or open-grown pines.

The term *heartwood* derives solely from its position and not from any vital importance to the tree. This is evidenced by the fact that a tree can thrive with its heart completely decayed. Some species begin to form heartwood very early in life, so having only a thin layer of live sapwood, while in others the change comes slowly. Thin sapwood is characteristic of such species as chestnut, black locust, mulberry, osage-orange, and sassafras, while in maple, ash, hickory, hackberry, beech, and pine, thick sapwood is the rule. Others never form heartwood.

No definite relation exists between the annual rings of growth and the amount of sapwood. Within the same species the cross-sectional area of the sapwood is very roughly proportional to the size of the crown of the tree. If the rings are narrow, more of them are required than where they are wide. As the tree gets larger, the sapwood must necessarily become thinner or increase materially in volume. Sapwood is thicker in the upper portion of the trunk of a tree than near the base, because the age and the diameter of the upper sections are less.

When a tree is very young it is covered with limbs almost, if not entirely, to the ground, but as it grows older some or all of them will eventually die and are either broken off or fall off. Subsequent growth of wood may completely conceal the stubs which will however remain as knots. No matter how smooth and clear a log is on the outside, it is more or less knotty near the middle. Consequently the sapwood of an old tree, and particularly of a forest-grown tree, will be freer from knots than the inner heartwood. Since in most uses of wood, knots are defects that weaken the timber and interfere with its ease of working and other properties, it follows that a given piece of sapwood, because of its position in the tree, may well be stronger than a piece of heartwood from the same tree.

It is remarkable that the inner heartwood of old trees remains as sound as it usually does, since in many cases it is hundreds, and in a few instances thousands, of years old. Every broken limb or root, or deep wound from fire, insects, or falling timber, may afford an entrance for decay, which, once started, may penetrate to all parts of the trunk. The larvae of many insects bore into the trees and their tunnels remain indefinitely as sources of weakness. Whatever advantages, however, that sapwood may have in this connection are due solely to its relative age and position.

If a tree grows all its life in the open and the conditions of soil and site remain unchanged, it will make its most rapid growth in youth, and gradually decline. The annual rings of growth are for many years quite wide, but later they become narrower and narrower. Since each succeeding ring is laid down on the outside of the wood previously formed, it follows that unless a tree materially increases its production of wood from year to year, the rings must necessarily become thinner as the trunk gets wider.

As a tree reaches maturity its crown becomes more open and the annual wood production is lessened, thereby reducing still more the width of the growth rings. In the case of forest-grown trees so much depends upon the competition of the trees in their struggle for light and nourishment that periods of rapid and slow growth may alternate. Some trees, such as southern oaks, maintain the same width of ring for hundreds of years. Upon the whole, however, as a tree gets larger in diameter the width of the growth rings decreases.

Different pieces of wood cut from a large tree may differ decidedly, particularly if the tree is big and mature. In some trees, the wood laid on late in the life of a tree is softer, lighter, weaker, and more even-textured than that produced earlier, but in other trees, the reverse applies. This may or may not correspond to heartwood and sapwood. In a large log the sapwood, because of the time in the life of the tree when it was grown, may be inferior in hardness, strength, and toughness to equally sound heartwood from the same log. In a smaller tree, the reverse may be true.

Colour

In species which show a distinct difference between heartwood and sapwood the natural colour of heartwood is usually darker than that of the sapwood, and very frequently the contrast is conspicuous. This is produced by deposits in the heartwood of chemical substances, so that a dramatic colour difference does not mean a dramatic difference in the mechanical properties of heartwood and sapwood, although there may be a dramatic chemical difference.

Some experiments on very resinous Longleaf Pine specimens indicate an increase in strength, due to the resin which increases the strength when dry. Such resin-saturated heartwood is called "fat lighter". Structures built of fat lighter are almost impervious to rot and termites; however they are very flammable. Stumps of old longleaf pines are often dug, split into small pieces and sold as kindling for fires. Stumps thus dug may actually remain a century or more since being cut. Spruce impregnated with crude resin and dried is also greatly increased in strength thereby.

Since the latewood of a growth ring is usually darker in colour than the earlywood, this fact may be used in judging the density, and therefore the hardness and strength of the material. This is particularly the case with coniferous woods. In ring-porous woods the vessels of the early wood not infrequently appear on a finished surface as darker than the denser latewood, though on cross sections of heartwood the reverse is commonly true. Except in the manner just stated the colour of wood is no indication of strength.

Abnormal discolouration of wood often denotes a diseased condition, indicating unsoundness. The black check in western hemlock is the result of insect attacks. The reddish-brown streaks so common in hickory and certain other woods are mostly the result of injury by birds. The discolouration is merely an indication of an injury, and in all probability does not of itself affect the properties of the wood. Certain rot-producing fungi impart to wood characteristic colours which thus become symptomatic of weakness; however an attractive effect known as spalting produced by this process is often considered a desirable characteristic. Ordinary sap-staining is due to fungal growth, but does not necessarily produce a weakening effect.

Water Content

Water occurs in living wood in three conditions, namely: (1) in the cell walls, (2) in the protoplasmic contents of the cells, and (3) as free water in the cell cavities and spaces. In heartwood it occurs only in the first and last forms. Wood that is thoroughly air-dried retains 8–16% of the water in the cell walls, and none, or practically none, in the other forms. Even oven-dried wood retains a small percentage of moisture, but for all except chemical purposes, may be considered absolutely dry.

The general effect of the water content upon the wood substance is to render it softer and more pliable. A similar effect of common observation is in the softening action of water on paper or cloth. Within certain limits, the greater the water content, the greater its softening effect.

Drying produces a decided increase in the strength of wood, particularly in small specimens. An extreme example is the case of a completely dry spruce block 5 cm in section, which will sustain a permanent load four times as great as a green (undried) block of the same size will.

The greatest strength increase due to drying is in the ultimate crushing strength, and strength at elastic limit in endwise compression; these are followed by the modulus of rupture, and stress at elastic limit in cross-bending, while the modulus of elasticity is least affected.

Structure

Wood is a heterogeneous, hygroscopic, cellular and anisotropic material. It is composed of cells, and the cell walls are composed of micro-fibrils of cellulose (40% – 50%) and hemicellulose (15% – 25%) impregnated with lignin (15% – 30%). In coniferous or softwood species the wood cells are mostly of one kind, tracheids, and as a result the material is much more uniform in structure than that of most hardwoods. There are no vessels ("pores") in coniferous wood such as one sees so prominently in oak and ash, for example. The structure of hardwoods is more complex. The water conducting capability is mostly taken care of by vessels: in some cases (oak, chestnut, ash) these are quite large and distinct, in others (buckeye, poplar, willow) too small to be seen without a hand lens. In discussing such woods it is customary to divide them into two large classes, *ring-porous* and *diffuse-porous*. In ring-porous species, such as ash, black locust, catalpa, chestnut, elm, hickory, mulberry, and oak, the larger vessels or pores (as cross sections of vessels are called) are localised in the part of the growth ring formed in spring, thus forming a region of more or less open and porous tissue. The rest of the ring, produced in summer, is made up of smaller vessels and a much greater proportion of wood fibres. These fibre are the elements which give strength and toughness to wood, while the vessels are a source of weakness.

In diffuse-porous woods the pores are evenly sized so that the water conducting capability is scattered throughout the growth ring instead of being collected in a band or row. Examples of this kind of wood are basswood, birch, buckeye, maple, poplar, and willow. Some species, such as walnut and cherry, are on the border between the two classes, forming an intermediate group.

Earlywood and Latewood in Softwood

In temperate softwoods there often is a marked difference between latewood and earlywood. The latewood will be denser than that formed early in the season. When examined under a microscope the cells of dense latewood are seen to be very thick-walled and with very small cell cavities, while those formed first in the season have thin walls and large cell cavities. The strength is in the walls, not the cavities. Hence the greater the proportion of latewood the greater the density and strength. In choosing a piece of pine where strength or stiffness is the important consideration, the principal thing to observe is the comparative amounts of earlywood and latewood. The width of ring is not nearly so important as the proportion and nature of the latewood in the ring.

If a heavy piece of pine is compared with a lightweight piece it will be seen at once that the heavier one contains a larger proportion of latewood than the other, and is therefore showing more clearly demarcated growth rings. In white pines there is not much contrast between the different parts of the ring, and as a result the wood is very uniform in texture and is easy to work. In hard pines, on the other hand, the latewood is very dense and is deep-coloured, presenting a very decided contrast to the soft, straw-coloured earlywood.

It is not only the proportion of latewood, but also its quality, that counts. In specimens that show a very large proportion of latewood it may be noticeably more porous and weigh considerably less than the latewood in pieces that contain but little. One can judge comparative density, and therefore to some extent strength, by visual inspection.

No satisfactory explanation can as yet be given for the exact mechanisms determining the formation of earlywood and latewood. Several factors may be involved. In conifers, at least, rate of growth alone does not determine the proportion of the two portions of the ring, for in some cases the wood of slow growth is very hard and heavy, while in others the opposite is true. The quality of the site where the tree grows undoubtedly affects the character of the wood formed, though it is not possible to formulate a rule governing it. In general, however, it may be said that where strength or ease of working is essential, woods of moderate to slow growth should be chosen.

Earlywood and Latewood in Ring-porous Woods

In ring-porous woods each season's growth is always well defined, because the large pores formed early in the season abut on the denser tissue of the year before.

In the case of the ring-porous hardwoods there seems to exist a pretty definite relation between the rate of growth of timber and its properties. This may be briefly summed up in the general statement that the more rapid the growth or the wider the rings of growth, the heavier, harder, stronger, and stiffer the wood. This, it must be remembered, applies only to ring-porous woods such as oak, ash, hickory, and others of the same group, and is, of course, subject to some exceptions and limitations.

In ring-porous woods of good growth it is usually the latewood in which the thick-walled, strength-giving fibres are most abundant. As the breadth of ring diminishes, this latewood is reduced so that very slow growth produces comparatively light, porous wood composed of thin-walled vessels and wood parenchyma. In good oak these large vessels of the earlywood occupy from 6 to 10 percent of the volume of the log, while in inferior material they may make up 25% or more. The latewood of good oak is dark coloured and firm, and consists mostly of thick-walled fibres which form one-half or more of the wood. In inferior oak, this latewood is much reduced both in quantity and quality. Such variation is very largely the result of rate of growth.

Wide-ringed wood is often called "second-growth", because the growth of the young timber in open stands after the old trees have been removed is more rapid than in trees in a closed forest, and in the manufacture of articles where strength is an important consideration such "second-growth" hardwood material is preferred. This is particularly the case in the choice of hickory for handles and spokes. Here not only strength, but toughness and resilience are important. The results of a series of tests on hickory by the U.S. Forest Service show that:

"The work or shock-resisting ability is greatest in wide-ringed wood that has from 5 to 14 rings per inch (rings 1.8-5 mm thick), is fairly constant from 14 to 38 rings per inch (rings 0.7–1.8 mm thick), and decreases rapidly from 38 to 47 rings per inch (rings 0.5–0.7 mm thick). The strength at maximum load is not so great with the most rapid-growing wood; it is maximum with from 14 to 20 rings per inch (rings 1.3–1.8 mm thick), and again becomes less as the wood becomes more closely ringed.

The natural deduction is that wood of first-class mechanical value shows from 5 to 20 rings per inch (rings 1.3–5 mm thick) and that slower growth yields poorer stock. Thus the inspector or buyer of hickory should discriminate against timber that has more than 20 rings per inch (rings less than 1.3 mm thick). Exceptions exist, however, in the case of normal growth upon dry situations, in which the slow-growing material may be strong and tough."

The effect of rate of growth on the qualities of chestnut wood is summarised by the same authority as follows:

"When the rings are wide, the transition from spring wood to summer wood is gradual, while in the narrow rings the spring wood passes into summer wood abruptly. The width of the spring wood changes but little with the width of the annual ring, so that the narrowing or broadening of the annual ring is always at the expense of the summer wood. The narrow vessels of the summer wood make it richer in wood substance than the spring wood composed of wide vessels.

Therefore, rapid-growing specimens with wide rings have more wood substance than slow-growing trees with narrow rings. Since the more the wood substance the greater the weight, and the greater the weight the stronger the wood, chestnuts with wide rings must have stronger wood than chestnuts with narrow rings. This agrees with the accepted view that sprouts (which always have wide rings) yield better and stronger wood than seedling chestnuts, which grow more slowly in diameter."

Earlywood and Latewood in Diffuse-porous Woods

In the diffuse-porous woods, the demarcation between rings is not always so clear and in some cases is almost (if not entirely) invisible to the unaided eye. Conversely, when there is a clear demarcation there may not be a noticeable difference in structure within the growth ring.

In diffuse-porous woods, as has been stated, the vessels or pores are even-sized, so that the water conducting capability is scattered throughout the ring instead of collected in the earlywood. The effect of rate of growth is, therefore, not the same as in the ring-porous woods, approaching more nearly the conditions in the conifers. In general it may be stated that such woods of medium growth afford stronger material than when very rapidly or very slowly grown.

In many uses of wood, total strength is not the main consideration. If ease of working is prized, wood should be chosen with regard to its uniformity of texture and straightness of grain, which will in most cases occur when there is little contrast between the latewood of one season's growth and the earlywood of the next.

Monocot Wood

Structural material that roughly (in its gross handling characteristics) resembles ordinary, "dicot" or conifer wood is produced by a number of monocot plants, and these also are colloquially called wood. Of these, bamboo, botanically a member of the grass family, has considerable economic importance, larger culms being widely used as a building and construction material in their own right and, these days, in the manufacture of engineered flooring, panels and veneer. Another major plant group that produce material that often is called wood are the palms. Of much less importance are plants such as *Pandanus, Dracaena* and *Cordyline.* With all this material, the structure and composition of the structural material is quite different from ordinary wood.

Hard and Soft Woods

There is a strong relationship between the properties of wood and the properties of the particular tree that yielded it. The density of wood varies with species. The density of a wood correlates with its strength (mechanical properties). For example, mahogany is a medium-dense hardwood that is excellent for fine furniture crafting, whereas balsa is light, making it useful for model building. One of the densest woods is black ironwood.

It is common to classify wood as either softwood or hardwood. The wood from conifers (e.g. pine) is called softwood, and the wood from dicotyledons (usually broad-leaved trees, e.g. oak) is called hardwood. These names are a bit misleading, as hardwoods are not necessarily hard, and softwoods are not necessarily soft. The well-known balsa (a hardwood) is actually softer than any commercial softwood. Conversely, some softwoods (e.g. yew) are harder than many hardwoods.

Chemistry of Wood

Aside from water, wood has three main components. Cellulose, a crystalline polymer derived from glucose, constitutes about 41–43%. Next in abundance is hemicellulose, which is around 20% in deciduous trees but near 30% in conifers. It is mainly five-carbon sugars that are linked in an irregular manner, in contrast to the cellulose. Lignin is the third component at around 27% in coniferous wood vs 23% in deciduous trees. Lignin confers the hydrophobic properties reflecting the fact that it is based on aromatic rings. These three components are interwoven, and direct covalent linkages exist between the lignin and the hemicellulose. A major focus of the paper industry is the separation of the lignin from the cellulose, from which paper is made. In chemical terms, the difference between hardwood and softwood is reflected in the composition of the constituent lignin. Hardwood lignin is primarily derived from sinapyl alcohol and coniferyl alcohol. Softwood lignin is mainly derived from coniferyl alcohol.

Extractives

Aside from the lignocellulose, wood consists of a variety of low molecular weight organic compounds, called *extractives*. The wood extractives are fatty acids, resin acids, waxes and terpenes. For example, rosin is exuded by conifers as protection from insects. The extraction of these organic materials from wood provides tall oil, terpentine, and rosin.

Uses

Fuel: Wood has a long history of being used as fuel, which continues to this day, mostly in rural areas of the world. Hardwood is preferred over softwood because it creates less smoke and burns longer. Adding a woodstove or fireplace to a home is often felt to add ambiance and warmth.

Construction

Wood has been an important construction material since humans began building shelters, houses and boats. Nearly all boats were made out of wood until the late 19th century, and wood remains in common use today in boat construction.

Wood to be used for construction work is commonly known as *lumber* in North America. Elsewhere, *lumber* usually refers to felled trees, and the word for sawn planks ready for use is *timber*.

New domestic housing in many parts of the world today is commonly made from timber-framed construction. Engineered wood products are becoming a bigger part of the construction industry. They may be used in both residential and commercial buildings as structural and aesthetic materials.

In buildings made of other materials, wood will still be found as a supporting material, especially in roof construction, in interior doors and their frames, and as exterior cladding. Wood is also commonly used as shuttering material to form the mould into which concrete is poured during reinforced concrete construction.

Furniture and Utensils

Wood has always been used extensively for furniture, such as chairs and beds. Also for tool handles and cutlery, such as chopsticks, toothpicks, and other utensils, like the wooden spoon.

Engineered Wood

Engineered wood products, glued building products "engineered" for application-specific performance requirements, are often used in construction and industrial applications. Glued engineered wood products are manufactured by bonding together wood strands, veneers, lumber or other forms of wood fibre with glue to form a larger, more efficient composite structural unit.

These products include glued laminated timber (glulam), wood structural panels (including plywood, oriented strand board and composite panels), laminated veneer lumber (LVL) and other structural composite lumber (SCL) products, parallel strand lumber, and I-joists. Approximately 100 million cubic meters of wood was consumed for this purpose in 1991. The trends suggest that particle board and fibre board will overtake plywood.

Engineered wood products display highly predictable and reliable performance characteristics and provide enhanced design flexibility: on one hand, these products allow the use of smaller pieces, and on the other hand, they allow for bigger spans. They may also be selected for specific projects such as public swimming pools or ice rinks where the wood will not deteriorate in the presence of certain chemicals, and are less susceptible to the humidity changes commonly found in these environments.

Engineered wood products prove to be more environmentally friendly and, if used appropriately, are often less expensive than building materials such as steel or concrete. These products are extremely resource-efficient because they use more of the available resource with minimal waste. In most cases, engineered wood products are produced using faster growing and often underutilized wood species from managed forests and tree farms.

Wood unsuitable for construction in its native form may be broken down mechanically (into fibres or chips) or chemically (into cellulose) and used as a raw material for other building materials, such as engineered wood, as well as chipboard, hardboard, and medium-density fibreboard (MDF).

Such wood derivatives are widely used: wood fibres are an important component of most paper, and cellulose is used as a component of some synthetic materials. Wood derivatives can also be used for kinds of flooring, for example laminate flooring.

Next Generation Wood Products

Further developments include new lignin glue applications, recyclable food packaging, rubber tire replacement applications, anti-bacterial medical agents, and high strength fabrics or composites. As scientists and engineers further learn and develop new techniques to extract various components from wood, or alternatively to modify wood, for example by adding components to wood, new more advanced products will appear on the marketplace.

In the Arts

Wood has long been used as an artistic medium. It has been used to make sculptures and carvings for millennia. Examples include the totem poles carved by North American indigenous people from conifer trunks, often Western Red Cedar (*Thuja plicata*), and the Millennium clock tower, now housed in the National Museum of Scotland in Edinburgh.

It is also used in woodcut printmaking, and for engraving. Certain types of musical instruments, such as those of the violin family, the guitar, the clarinet and recorder, the xylophone, and the marimba, are made mostly or entirely of wood. The choice of wood may make a significant difference to the tone and resonant qualities of the instrument, and tonewoods have widely differing properties, ranging from the hard and dense african blackwood (used for the bodies of clarinets) to the light but resonant European spruce (*Picea abies*) (traditionally used for the soundboards of violins). The most valuable tonewoods, such as the ripple sycamore (*Acer pseudoplatanus*), used for the backs of violins, combine acoustic properties with decorative colour and grain which enhance the appearance of the finished instrument.

Despite their collective name, not all woodwind instruments are made entirely of wood. The reeds used to play them, however, are usually made from *Arundo donax*, a type of monocot cane plant.

Sports and Recreational Equipment

Many types of sports equipment are made of wood, or were constructed of wood in the past. For example, cricket bats are typically made of white willow. The baseball bats which are legal for use in Major League Baseball are frequently made of ash wood or hickory, and in recent years have been constructed from maple even though that wood is somewhat more fragile. In softball, however, bats are more commonly made of aluminium (this is especially true for fastpitch softball).

Many other types of sports and recreation equipment, such as skis, ice hockey sticks, lacrosse sticks and archery bows, were commonly made of wood in the past, but have since been replaced with more modern materials such as aluminium, fibreglass, carbon fibre, titanium, and composite materials. One noteworthy example of this trend is the golf club commonly known as the *wood*, the head of which was traditionally made of persimmon wood in the early days of the game of golf, but is now generally made of synthetic materials.

Medicine

In January 2010 Italian scientists announced that wood could be harnessed to become a bone substitute. It is likely to take at least five years until this technique will be applied for humans.

Leaf

A leaf is an organ of a vascular plant, as defined in botanical terms, and in particular in plant morphology. Foliage is a mass noun that refers to leaves as a feature of plants.

Figure: *The leaves of a Beech tree*

Typically a leaf is a thin, flattened organ borne above ground and specialized for photosynthesis, but many types of leaves are adapted in ways almost unrecognisable in those terms: some are not flat (for example many succulent leaves and conifers), some are not above ground (such as bulb scales), and some are without major photosynthetic function (consider for example cataphylls, spines, and cotyledons).

Conversely, many structures of non-vascular plants, or even of some lichens, which are not plants at all (in the sense of being members of the kingdom Plantae), do look and function much like leaves. Furthermore, several structures found in vascular plants look like leaves but are not actually leaves; they differ from leaves in their structures and origins. Examples include phyllodes, cladodes, and phylloclades.

General Nature of Leaves

Typically leaves are flat and thin, thereby maximising the surface area directly exposed to light and promoting photosynthetic function. Externally they commonly are arranged on the plant in such ways as to expose their surfaces to light as efficiently as possible without shading each other, but there are many exceptions and complications; for instance plants adapted to windy conditions may have pendent leaves, such as in many willows and *Eucalyptus*.

Likewise, the internal organisation of most kinds of leaves has evolved to maximise exposure of the photosynthetic organelles, the chloroplasts, to light and to increase the absorption of carbon dioxide. Most leaves have stomata, which open or narrow to regulate the exchange of carbon dioxide, oxygen, and water vapour with the atmosphere.

In contrast however, some leaf forms are adapted to modulate the amount of light they absorb to avoid or mitigate excessive heat, ultraviolet damage, or desiccation, or to sacrifice light-absorption efficiency in favour of protection from herbivorous enemies. Among these forms the leaves of many xerophytes are conspicuous. For such plants their major constraint is not light flux or intensity, but heat, cold, drought, wind, herbivory, and various other hazards. Typical examples among such strategies are so-called window plants such as Fenestraria species, some Haworthia species such as Haworthia tesselata and Haworthia truncata and Bulbine mesembryanthemoides.

The shape and structure of leaves vary considerably from species to species of plant, depending largely on their adaptation to climate and available light, but also to other factors such as grazing animals, available nutrients, and ecological competition from other plants. Considerable changes in leaf type occur within species too, for example as a plant matures; as a case in point *Eucalyptus* species commonly have isobilateral, pendent leaves when mature and dominating their neighbours; however, such trees tend to have erect or horizontal dorsiventral leaves as seedlings, when their growth is limited by the available light. Other factors include the need to balance water loss at high temperature and low humidity against the need to absorb atmospheric carbon dioxide. In most plants leaves also are the primary organs responsible for transpiration and guttation (beads of fluid forming at leaf margins).

Leaves can also store food and water, and are modified accordingly to meet these functions, for example in the leaves of succulent plants and in bulb scales. The concentration of photosynthetic structures in leaves requires that they be richer in protein, minerals, and sugars, than say, woody stem tissues. Accordingly leaves are prominent in the diet of many animals. This is true for humans, for whom leaf vegetables commonly are food staples.

Figure: *A Leaf Shed in Autumn.*

Correspondingly, leaves represent heavy investment on the part of the plants bearing them, and their retention or disposition are the subject of elaborate strategies for dealing with pest pressures, seasonal conditions, and protective measures such as the growth of thorns and the production of phytoliths, lignins, tannins and poisons.

Deciduous plants in frigid or cold temperate regions typically shed their leaves in autumn, whereas in areas with a severe dry season, some plants may shed their leaves until the dry season ends. In either case the shed leaves may be expected to contribute their retained nutrients to the soil where they fall.

In contrast, many other non-seasonal plants, such as palms and conifers, retain their leaves for long periods; *Welwitschia* retains its two main leaves throughout a lifetime that may exceed a thousand years.

Not all plants have true leaves. Bryophytes (e.g., mosses and liverworts) are non-vascular plants, and, although they produce flattened, leaf-like structures that are rich in chlorophyll, these organs differ morphologically from the leaves of vascular plants; For one thing, they lack vascular tissue. Vascularised leaves first evolved following the Devonian period, when carbon dioxide concentration in the atmosphere dropped significantly. This occurred independently in two separate lineages of vascular plants: the microphylls of lycophytes and the euphylls ("true leaves") of ferns, gymnosperms, and angiosperms. Euphylls are also referred to as macrophylls or megaphylls ("large leaves").

Large-scale Features (Leaf Morphology)

A structurally complete leaf of an angiosperm consists of a petiole (leaf stalk), a lamina (leaf blade), and stipules (small structures located to either side of the base of the petiole). Not every species produces leaves with all of these structural components. In certain species, paired stipules are not obvious or are absent altogether. A petiole may be absent, or the blade may not be laminar (flattened). The tremendous variety shown in leaf structure (anatomy) from species to species is presented in detail below under morphology.

The petiole mechanically links the leaf to the plant and provides the route for transfer of water and sugars to and from the leaf. The lamina is typically the location of the majority of photosynthesis.

Anatomy

Medium Scale Features: Leaves are normally extensively vascularised and are typically covered by a dense network of xylem, which supply water for photosynthesis, and phloem, which remove the sugars produced by photosynthesis. Many leaves are covered in trichomes (small hairs) which have a diverse range of structures and functions.

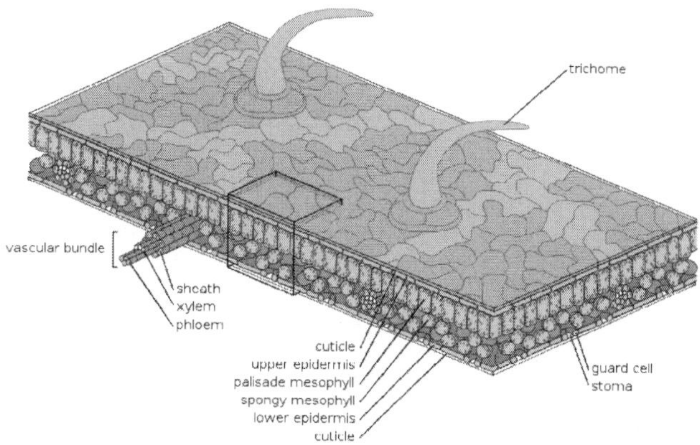

Small-scale Features

A leaf is a plant organ and is a collection of tissues in a regular organisation. The major tissue systems present are

1. The epidermis, which covers the upper and lower surfaces
2. The mesophyll inside the leaf, which is rich in chloroplasts (also called chlorenchyma)
3. The arrangement of veins (the vascular tissue).

These three tissue systems typically form a regular organisation at the cellular scale.

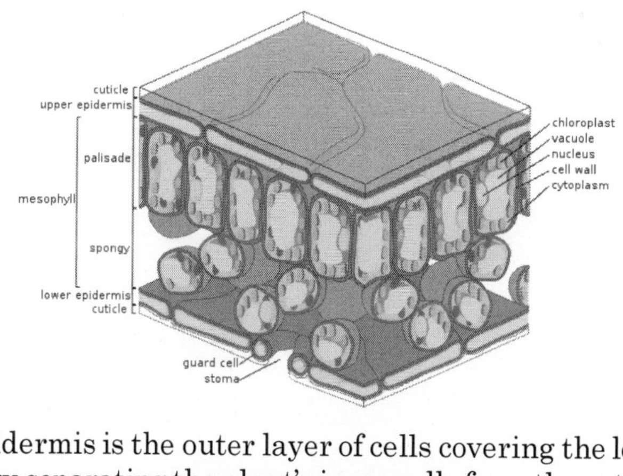

The epidermis is the outer layer of cells covering the leaf. It forms the boundary separating the plant's inner cells from the external world. The epidermis serves several functions: protection against water loss by way of transpiration, regulation of gas exchange, secretion of metabolic compounds, and (in some species) absorption of water. Most leaves show dorsoventral anatomy: The upper (adaxial) and lower (abaxial) surfaces have somewhat different construction and may serve different functions.

The epidermis is usually transparent (epidermal cells lack chloroplasts) and coated on the outer side with a waxy cuticle that prevents water loss. The cuticle is in some cases thinner on the lower epidermis than on the upper epidermis, and is generally thicker on leaves from dry climates as compared with those from wet climates.

Epidermis

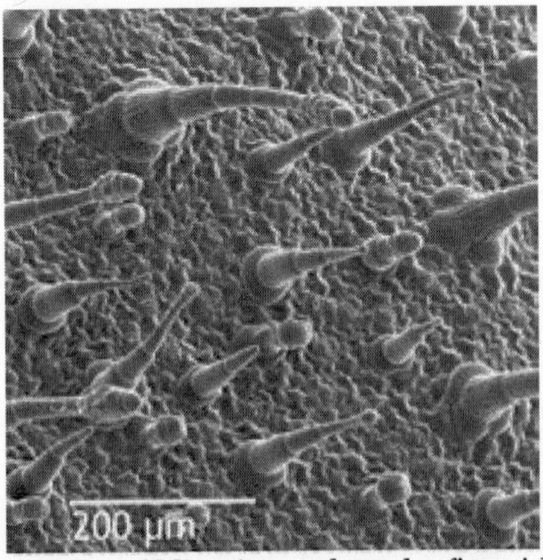

Figure: *SEM image of* Nicotiana alata *leaf's epidermis, showing trichomes (hair-like appendages) and stomata (eye-shaped slits, visible at full resolution).*

The epidermis tissue includes several differentiated cell types: epidermal cells, epidermal hair cells (trichomes) cells in the stomate complex; guard cells and subsidiary cells. The epidermal cells are the most numerous, largest, and least specialized and form the majority of the epidermis. These are typically more elongated in the leaves of monocots than in those of dicots.

The epidermis is covered with pores called *stomata,* part of a stoma complex consisting of a pore surrounded on each side by chloroplast-containing guard cells, and two to four subsidiary cells that lack chloroplasts. Opening and closing of the stoma complex regulates the exchange of gases and water vapour between the outside air and the interior of the leaf and plays an important role in allowing photosynthesis without letting the leaf dry out. In a typical leaf, the stomata are more numerous over the abaxial (lower) epidermis than the adaxial (upper) epidermis and more numerous in plants from cooler climates.

Mesophyll

Most of the interior of the leaf between the upper and lower layers of epidermis is a *parenchyma* (ground tissue) or *chlorenchyma* tissue called the mesophyll (Greek for "middle leaf"). This assimilation tissue is the primary location of photosynthesis in the plant. The products of photosynthesis are called "assimilates".

In ferns and most flowering plants, the mesophyll is divided into two layers:

- An upper palisade layer of tightly packed, vertically elongated cells, one to two cells thick, directly beneath the adaxial epidermis. Its cells contain many more chloroplasts than the spongy layer. These long cylindrical cells are regularly arranged in one to five rows. Cylindrical cells, with the *chloroplasts* close to the walls of the cell, can take optimal advantage of light. The slight separation of the cells provides maximum absorption of carbon dioxide. This separation must be minimal to afford capillary action for water distribution. In order to adapt to their different environment (such as sun or shade), plants had to adapt this structure to obtain optimal result. Sun leaves have a multi-layered palisade layer, while shade leaves or older leaves closer to the soil are single-layered.
- Beneath the palisade layer is the spongy layer. The cells of the spongy layer are more rounded and not so tightly packed. There are large intercellular air spaces. These cells contain fewer chloroplasts than those of the palisade layer. The pores or *stomata* of the epidermis open into substomatal chambers, which are connected to the air spaces between the spongy layer cells.

These two distinct layers of the mesophyll are absent in many aquatic and marsh plants. Even an epidermis and a mesophyll may be lacking. Instead, for their gaseous exchanges they use a homogeneous aerenchyma (thin-walled cells separated by large gas-filled spaces). Their stomata are situated at the upper surface.

Leaves are normally green, due to chlorophyll in plastids in the chlorenchyma cells. Plants that lack chlorophyll cannot photosynthesize.

Veins

The veins are the vascular tissue of the leaf and are located in the spongy layer of the mesophyll. They are typical examples of pattern formation through ramification. The pattern of the veins is called venation.

The veins are made up of:

- *Xylem:* tubes that bring water and minerals from the roots into the leaf.
- *Phloem:* tubes that usually move sap, with dissolved sucrose, produced by photosynthesis in the leaf, out of the leaf.

The xylem typically lies on the adaxial side of the vascular bundle and the phloem typically lies on the abaxial side. Both are embedded in a dense parenchyma tissue, called the pith or sheath, which usually includes some structural collenchyma tissue.

Seasonal Leaf Loss

Leaves in temperate, boreal, and seasonally dry zones may be seasonally deciduous (falling off or dying for the inclement season). This mechanism to shed leaves is called abscission. After the leaf is shed, a leaf scar develops on the twig. In cold autumns, they sometimes change colour, and turn yellow, bright-orange, or red, as various accessory pigments (carotenoids and xanthophylls) are revealed when the tree responds to cold and reduced sunlight by curtailing chlorophyll production. Red anthocyanin pigments are now thought to be produced in the leaf as it dies, possibly to mask the yellow hue left when the chlorophyll is lost - yellow leaves appear to attract herbivores such as aphids.

Morphology

External leaf characteristics (such as shape, margin, hairs, etc.) are important for identifying plant species, and botanists have developed a rich terminology for describing leaf characteristics. These structures are a part of what makes leaves determinant; they grow and achieve a specific pattern and shape, then stop. Other plant parts like stems or roots are non-determinant, and will usually continue to grow as long as they have the resources to do so.

Classification of leaves can occur through many different designative schema, and the type of leaf is usually characteristic of a species, although some species produce more than one type of leaf. The longest type of leaf is a leaf from a palm, measuring at nine feet long. The terminology associated with the description of leaf morphology is presented, in illustrated form, at Wikibooks.

Basic Types

- Ferns have fronds
- Conifer leaves are typically needle-, awl-, or scale-shaped

- Angiosperm (flowering plant) leaves: the standard form includes stipules, a petiole, and a lamina
- Lycophytes have microphyll leaves.
- Sheath leaves (type found in most grasses)
- Other specialized leaves (such as those of *Nepenthes*)

Arrangement on the Stem

Different terms are usually used to describe leaf placement (phyllotaxis):

Figure: *The leaves on this plant are arranged in pairs opposite one another, with successive pairs at right angles to each other ("decussate") along the red stem. Note the developing buds in the axils of these leaves.*

- Alternate – leaf attachments are singular at nodes, and leaves alternate direction, to a greater or lesser degree, along the stem.
- Opposite – Two structures, one on each opposite side of the stem, typically leaves, branches, or flower parts. Leaf attachments are paired at each node; decussate if, as typical, each successive pair is rotated 90° progressing along the stem; or distichous if not rotated, but two-ranked (in the same geometric flat-plane).
- Whorled – three or more leaves attach at each point or node on the stem. As with opposite leaves, successive whorls may or may not be decussate, rotated by half the angle between the

leaves in the whorl (i.e., successive whorls of three rotated 60°, whorls of four rotated 45°, etc.). Opposite leaves may appear whorled near the tip of the stem.

- Rosulate – leaves form a rosette

As a *stem* grows, leaves tend to appear arranged around the stem in a way that optimizes yield of light. In essence, leaves form a helix pattern centreed around the stem, either clockwise or counterclockwise, with (depending upon the species) the same angle of divergence. There is a regularity in these angles and they follow the numbers in a Fibonacci sequence: 1/2, 2/3, 3/5, 5/8, 8/13, 13/21, 21/34, 34/55, 55/89. This series tends to a limit close to 360° x 34/89 = 137.52 or 137° 30', an angle known in mathematics as the golden angle. In the series, the numerator indicates the number of complete turns or "gyres" until a leaf arrives at the initial position. The denominator indicates the number of leaves in the arrangement. This can be demonstrated by the following:

- alternate leaves have an angle of 180° (or 1/2)
- 120° (or 1/3) : three leaves in one circle
- 144° (or 2/5) : five leaves in two gyres
- 135° (or 3/8) : eight leaves in three gyres.

Divisions of the Blade

Two basic forms of leaves can be described considering the way the blade (lamina) is divided. A simple leaf has an undivided blade. However, the leaf shape may be formed of lobes, but the gaps between lobes do not reach to the main vein. A compound leaf has a fully subdivided blade, each leaflet of the blade separated along a main or secondary vein. Because each leaflet can appear to be a simple leaf, it is important to recognize where the petiole occurs to identify a compound leaf. Compound leaves are a characteristic of some families of higher plants, such as the Fabaceae. The middle vein of a compound leaf or a frond, when it is present, is called a rachis.

- *Palmately compound* leaves have the leaflets radiating from the end of the petiole, like fingers off the palm of a hand, e.g. *Cannabis* (hemp) and *Aesculus* (buckeyes).
- *Pinnately compound* leaves have the leaflets arranged along the main or mid-vein.
    - o odd pinnate: with a terminal leaflet, e.g. *Fraxinus* (ash).
    - o even pinnate: lacking a terminal leaflet, e.g. *Swietenia* (mahogany).

- *Bipinnately compound* leaves are twice divided: the leaflets are arranged along a secondary vein that is one of several branching off the rachis. Each leaflet is called a "pinnule". The pinnules on one secondary vein are called "pinna"; e.g. *Albizia* (silk tree).
- *trifoliate* (or *trifoliolate*): a pinnate leaf with just three leaflets, e.g. *Trifolium* (clover), *Laburnum* (laburnum).
- *pinnatifid*: pinnately dissected to the central vein, but with the leaflets not entirely separate, e.g. *Polypodium*, some *Sorbus* (whitebeams). In pinnately veined leaves the central vein in known as the *midrib*.

Characteristics of the Petiole

Petiolated leaves have a petiole (leaf stem). Sessile leaves do not: The blade attaches directly to the stem. In clasping or decurrent leaves, the blade partially or wholly surrounds the stem, often giving the impression that the shoot grows through the leaf. When this is the case, the leaves are called "perfoliate", such as in *Claytonia perfoliata*. In peltate leaves, the petiole attaches to the blade inside from the blade margin.

In some *Acacia* species, such as the Koa Tree (*Acacia koa*), the petioles are expanded or broadened and function like leaf blades; these are called phyllodes. There may or may not be normal pinnate leaves at the tip of the phyllode.

A stipule, present on the leaves of many dicotyledons, is an appendage on each side at the base of the petiole resembling a small leaf. Stipules may be lasting and not be shed (a stipulate leaf, such as in roses and beans), or be shed as the leaf expands, leaving a stipule scar on the twig (an exstipulate leaf).

- The situation, arrangement, and structure of the stipules is called the "stipulation".
  - o free
  - o adnate : fused to the petiole base
  - o ochreate : provided with ochrea, or sheath-formed stipules, e.g. rhubarb,
  - o encircling the petiole base
  - o interpetiolar : between the petioles of two opposite leaves.
  - o intrapetiolar : between the petiole and the subtending stem.

Venation

There are two subtypes of venation, namely, *craspedodromous*, where the major veins stretch up to the margin of the leaf, and *camptodromous*, when major veins extend close to the margin, but bend before they intersect with the margin.

- Feather-veined, reticulate (also called pinnate-netted, penniribbed, penninerved, or penniveined) – the veins arise pinnately from a single mid-vein and subdivide into veinlets. These, in turn, form a complicated network. This type of venation is typical for (but by no means limited to) dicotyledons.
- Three main veins branch at the base of the lamina and run essentially parallel subsequently, as in *Ceanothus*. A similar pattern (with 3-7 veins) is especially conspicuous in Melastomataceae.
- Palmate-netted, palmate-veined, fan-veined; several main veins diverge from near the leaf base where the petiole attaches, and radiate towards the edge of the leaf, e.g. most *Acer* (maples).
- Parallel-veined, parallel-ribbed, parallel-nerved, penniparallel – veins run parallel for the length of the leaf, from the base to the apex. Commissural veins (small veins) connect the major parallel veins. Typical for most monocotyledons, such as grasses.
- Dichotomous – There are no dominant bundles, with the veins forking regularly by pairs; found in *Ginkgo* and some pteridophytes.

Note that, although it is the more complex pattern, branching veins appear to be plesiomorphic and in some form were present in ancient seed plants as long as 250 million years ago. A pseudo-reticulate venation that is actually a highly modified penniparallel one is an autapomorphy of some Melanthiaceae, which are monocots, e.g. *Paris quadrifolia* (True-lover's Knot).

Morphology Changes Within a Single Plant

- Homoblasty - Characteristic in which a plant has small changes in leaf size, shape, and growth habit between juvenile and adult stages.
- Heteroblasty - Characteristic in which a plant has marked changes in leaf size, shape, and growth habit between juvenile and adult stages.

Edge (Margin)

- *ciliate:* fringed with hairs
- *crenate:* wavy-toothed; dentate with rounded teeth, such as *Fagus* (beech)
- *crenulate* finely or shallowly crenate
- *dentate:* toothed, such as *Castanea* (chestnut)
    - o *coarse-toothed:* with large teeth
    - o *glandular toothed:* with teeth that bear glands.
- *denticulate:* finely toothed
- *doubly toothed:* each tooth bearing smaller teeth, such as *Ulmus* (elm)
- *entire:* even; with a smooth margin; without toothing
- *lobate:* indented, with the indentations not reaching to the centre, such as many *Quercus* (oaks)
    - o *palmately lobed:* indented with the indentations reaching to the centre, such as *Humulus* (hop).
- *serrate:* saw-toothed with asymmetrical teeth pointing forward, such as *Urtica* (nettle)
- *serrulate:* finely serrate
- *sinuate:* with deep, wave-like indentations; coarsely crenate, such as many *Rumex* (docks)
- *spiny* or *pungent:* with stiff, sharp points, such as some *Ilex* (hollies) and *Cirsium* (thistles).

Tip

- *acuminate:* long-pointed, prolonged into a narrow, tapering point in a concave manner.
- *acute:* ending in a sharp, but not prolonged point
- *cuspidate:* with a sharp, elongated, rigid tip; tipped with a cusp.
- *emarginate:* indented, with a shallow notch at the tip.
- *mucronate:* abruptly tipped with a small short point, as a continuation of the midrib; tipped with a mucro.
- *mucronulate:* mucronate, but with a smaller spine.
- *obcordate:* inversely heart-shaped, deeply notched at the top.
- *obtuse:* rounded or blunt
- *truncate:* ending abruptly with a flat end, that looks cut off.

Base

- *acuminate:* coming to a sharp, narrow, prolonged point.
- *acute:* coming to a sharp, but not prolonged point.
- *auriculate:* ear-shaped.
- *cordate:* heart-shaped with the notch towards the stalk.
- *cuneate:* wedge-shaped.
- *hastate:* shaped like an halberd and with the basal lobes pointing outward.
- *oblique:* slanting.
- *reniform:* kidney-shaped but rounder and broader than long.
- *rounded:* curving shape.
- *sagittate:* shaped like an arrowhead and with the acute basal lobes pointing downward.
- *truncate:* ending abruptly with a flat end, that looks cut off.

Surface

- *farinose:* bearing farina; mealy, covered with a waxy, whitish powder.
- *glabrous:* smooth, not hairy.
- *glaucous:* with a whitish bloom; covered with a very fine, bluish-white powder.
- *glutinous:* sticky, viscid.
- *papillate*, or *papillose:* bearing papillae (minute, nipple-shaped protuberances).
- *pubescent:* covered with erect hairs (especially soft and short ones).
- *punctate:* marked with dots; dotted with depressions or with translucent glands or coloured dots.
- *rugose:* deeply wrinkled; with veins clearly visible.
- *scurfy:* covered with tiny, broad scalelike particles.
- *tuberculate:* covered with tubercles; covered with warty prominences.
- *verrucose:* warted, with warty outgrowths.
- *viscid*, or *viscous:* covered with thick, sticky secretions.

The leaf surface is also host to a large variety of microorganisms; in this context it is referred to as the phyllosphere.

Hairiness

Figure: *Common Mullein (*Verbascum thapsus*) leaves are covered in dense, stellate trichomes.*

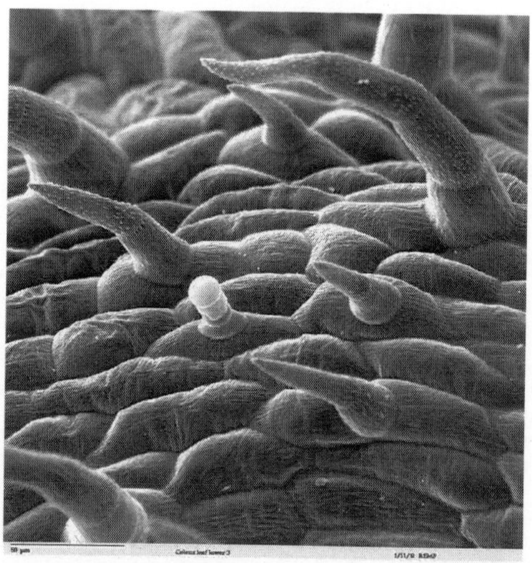

Figure: *Scanning electron microscope image of trichomes on the lower surface of a* Coleus blumei *(coleus) leaf.*

"Hairs" on plants are properly called trichomes. Leaves can show several degrees of hairiness. The meaning of several of the following terms can overlap.

- *arachnoid*, or *arachnose:* with many fine, entangled hairs giving a cobwebby appearance.
- *barbellate:* with finely barbed hairs (barbellae).

- *bearded:* with long, stiff hairs.
- *bristly:* with stiff hair-like prickles.
- *canescent:* hoary with dense grayish-white pubescence.
- *ciliate:* marginally fringed with short hairs (cilia).
- *ciliolate:* minutely ciliate.
- *floccose:* with flocks of soft, woolly hairs, which tend to rub off.
- *glabrescent:* losing hairs with age.
- *glabrous:* no hairs of any kind present.
- *glandular:* with a gland at the tip of the hair.
- *hirsute:* with rather rough or stiff hairs.
- *hispid:* with rigid, bristly hairs.
- *hispidulous:* minutely hispid.
- *hoary:* with a fine, close grayish-white pubescence.
- *lanate*, or *lanose:* with woolly hairs.
- *pilose:* with soft, clearly separated hairs.
- *puberulent*, or *puberulous:* with fine, minute hairs.
- *pubescent:* with soft, short and erect hairs.
- *scabrous*, or *scabrid:* rough to the touch.
- *sericeous:* silky appearance through fine, straight and appressed (lying close and flat) hairs.
- *silky:* with adpressed, soft and straight pubescence.
- *stellate*, or *stelliform:* with star-shaped hairs.
- *strigose:* with appressed, sharp, straight and stiff hairs.
- *tomentose:* densely pubescent with matted, soft white woolly hairs.
    - o *cano-tomentose:* between canescent and tomentose.
    - o *felted-tomentose:* woolly and matted with curly hairs.
- *tomentulose:* minutely or only slightly tomentose.
- *villous:* with long and soft hairs, usually curved.
- *woolly:' with long, soft and tortuous or matted hairs.*

Adaptations

In the course of evolution, leaves have adapted to different environments in the following ways:

- A certain surface structure avoids moistening by rain and contamination.

- Sliced leaves reduce wind resistance.
- Hairs on the leaf surface trap humidity in dry climates and create a boundary layer reducing water loss.
- Waxy leaf surfaces reduce water loss.
- Large surface area provides large area for sunlight and shade for plant to minimize heating and reduce water loss.
- In more or less opaque or buried in the soil leaves, translucent windows filter the light before the photosynthesis takes place at the inner leaf surfaces (e.g. *Fenestraria*).
- Succulent leaves store water and organic acids for use in CAM photosynthesis.
- Aromatic oils, poisons or pheromones produced by leaf borne glands deter herbivores (e.g. eucalypts).
- Inclusions of crystalline minerals deter herbivores (e.g. silica phytoliths in grasses, raphides in Araceae).
- Petals attracts pollinators.
- Spines protect the plants (*e.g.* cacti).
- Insect traps feed the plants directly.
- Bulbs store food and water (*e.g.* onions).
- Tendrils allow the plant to climb (*e.g.* peas).
- Bracts and pseudanthia (*false flowers*) replace normal flower structures when the true flowers are greatly reduced (*e.g.* Spurges).

Interactions with Other Organisms

Although not as nutritious as other organs such as fruit, leaves provide a food source for many organisms. Animals which eat leaves are known as folivores. The leaf is one of the most vital parts of the plant, and plants have evolved protection against folivores such as tannins, chemicals which hinder the digestion of proteins and have an unpleasant taste. Some animals have cryptic adaptations to avoid their own predators. For example, some caterpillars will create a small home in the leaf by folding it over themselves, while other herbivores and their prey mimic the appearance of the leaf. Some insects, such as the katydid, take this even further, moving from side to side much like a leaf does in the wind.

Meristem

A meristem is the tissue in most plants consisting of undifferentiated cells (meristematic cells), found in zones of the plant where growth can take place. The meristematic cells give rise to various organs of the plant, and keep the plant growing. The Shoot Apical Meristem (SAM) gives rise to organs like the leaves and flowers. The cells of the apical meristems - SAM and RAM (Root Apical Meristem) - divide rapidly and are considered to be indeterminate, in that they do not possess any defined end fate. In that sense, the meristematic cells are frequently compared to the stem cells in animals, that have an analogous behaviour and function.

The term *meristem* was first used in 1858 by Karl Wilhelm von Nägeli (1817–1891) in his book *Beiträge zur Wissenschaftlichen Botanik*. It is derived from the Greek word *merizein*, meaning to divide, in recognition of its inherent function.

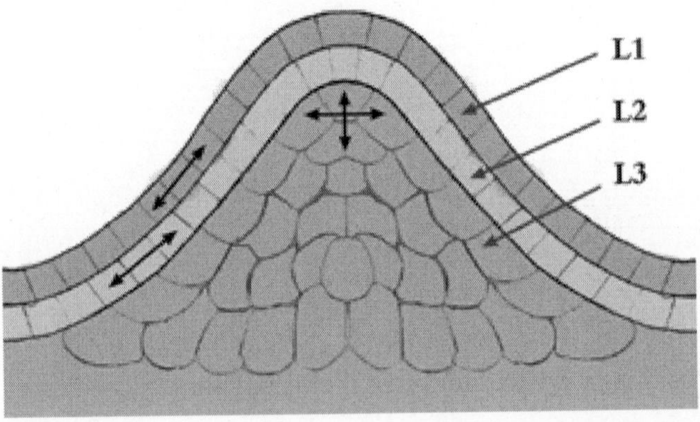

Figure: *Tunica-Corpus model of the apical meristem (growing tip). The epidermal (L1) and subepidermal (L2) layers form the outer layers called the tunica. The inner L3 layer is called the corpus. Cells in the L1 and L2 layers divide in a sideways fashion, which keeps these layers distinct, whereas the L3 layer divides in a more random fashion.*

In general, differentiated plant cells cannot divide or produce cells of a different type. Therefore, cell division in the meristem is required to provide new cells for expansion and differentiation of tissues and initiation of new organs, providing the basic structure of the plant body.

Meristematic cells are incompletely or not at all differentiated, and are capable of continued cellular division (youthful). Furthermore, the cells are small and protoplasm fills the cell completely. The vacuoles are extremely small. The cytoplasm does not contain differentiated plastids (chloroplasts or chromoplasts), although they are present in rudimentary form (proplastids). Meristematic cells are packed closely together without intercellular cavities. The cell wall is a very thin *primary cell wall*.

Maintenance of the cells requires a balance between two antagonistic processes: organ initiation and stem cell population renewal.

Meristematic Zones

Apical meristems are the completely undifferentiated (indeterminate) meristems in a plant. These differentiate into three kinds of primary meristems. The primary meristems in turn produce the two secondary meristem types. These secondary meristems are also known as lateral meristems because they are involved in lateral growth.

At the meristem summit, there is a small group of slowly dividing cells, which is commonly called the central zone. Cells of this zone have a stem cell function and are essential for meristem maintenance. The proliferation and growth rates at the meristem summit usually differ considerably from those at the periphery.

Meristems also are induced in the roots of legumes such as soybean, *Lotus japonicus*, pea, and *Medicago truncatula* after infection with soil bacteria commonly called Rhizobium. Cells of the inner or outer cortex in the so-called "window of nodulation" just behind the developing root tip are induced to divide. The critical signal substance is the lipo-oligosaccharide Nod-factor, decorated with side groups to allow specificity of interaction. The Nod factor receptor proteins NFR1 and NFR5 were cloned from several legumes including *Lotus japonicus*, *Medicago truncatula* and soybean (Glycine max). Regulation of nodule meristems utilizes long distance regulation commonly called "Autoregulation of Nodulation" (AON). This process involves a leaf-vascular tissue located LRR receptor kinases (LjHAR1, GmNARK and MtSUNN), CLE peptide signalling, and KAPP interaction, similar to that seen in the CLV1,2,3 system. LjKLAVIER also exhibits a nodule regulation phenotype though it is not yet known how this relates to the other AON receptor kinases

Apical Meristems

The apical meristem, or growing tip, is a completely undifferentiated meristematic tissue found in the buds and growing tips of roots in plants. Its main function is to begin growth of new cells in young seedlings at the tips of roots and shoots (forming buds, among other things). Specifically, an active apical meristem lays down a growing root or shoot behind itself, pushing itself forward. Apical meristems are very small, compared to the cylinder-shaped lateral meristems.

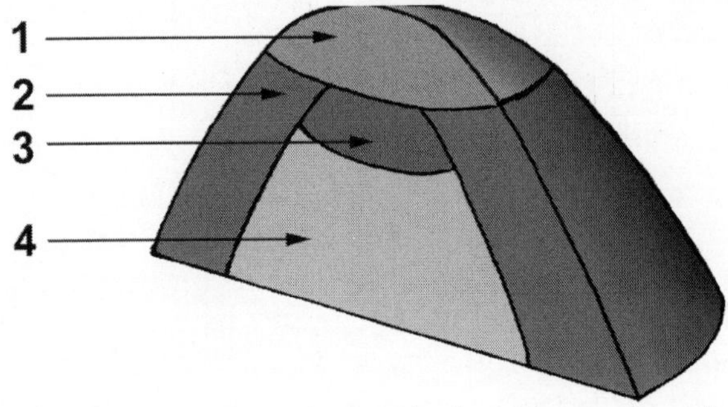

Figure: *Organisation of an apical meristem (growing tip)*

1 - Central zone

2 - Peripheral zone

3 - Medullary (i.e. central) meristem

4 - Medullary tissue

Apical meristems are composed of several layers. The number of layers varies according to plant type. In general the outermost layer is called the tunica while the innermost layers are the corpus. In monocots, the tunica determine the physical characteristics of the leaf edge and margin. In dicots, layer two of the corpus determine the characteristics of the edge of the leaf. The corpus and tunica play a critical part of the plant physical appearance as all plant cells are formed from the meristems. Apical meristems are found in two locations: the root and the stem. Some Arctic plants have an apical meristem in the lower/ middle parts of the plant. It is thought that this kind of meristem evolved because it is advantageous in Arctic conditions.

Shoot apical meristems

The source of all above-ground organs. Cells at the shoot apical meristem summit serve as stem cells to the surrounding peripheral region, where they proliferate rapidly and are incorporated into differentiating leaf or flower primordia.

The shoot apical meristem is the site of most of the embryogenesis in flowering plants. Primordia of leaves, sepals, petals, stamens and ovaries are initiated here at the rate of one every time interval, called a plastochron. It is where the first indications that flower development has been evoked are manifested. One of these indications might be the loss of apical dominance and the release of otherwise dormant cells to develop as axillary shoot meristems, in some species in axils of primordia as close as two or three away from the apical dome. The shoot apical meristem consists of 4 distinct cell groups: -.

- Stem cells
- The immediate daughter cells of the stem cells
- A subjacent organising centre
- Founder cells for organ initiation in surrounding regions

The four distinct zones mentioned above are maintained by a complex signalling pathway. In *Arabidopsis thaliana*, 3 interacting *CLAVATA* genes are required to regulate the size of the stem cell reservoir in the shoot apical meristem by controlling the rate of cell division. CLV1 and CLV2 are predicted to form a receptor complex (of the LRR receptor like kinase family) to which CLV3 is a ligand. CLV3 shares some homology with the ESR proteins of maize, with a short 14 amino acid region being conserved between the proteins. Proteins that contain these conserved regions have been grouped into the CLE family of proteins.

CLV1 has been shown to interact with several cytoplasmic proteins that are most likely involved in downstream signalling. For example, the CLV complex has been found to be associated with Rho/Rac small GTPase-related proteins.

These proteins may act as an intermediate between the CLV complex and a mitogen-activated protein kinase (MAPK), which is often involved in signalling cascades. KAPP is a kinase-associated protein phosphatase that has been shown to interact with CLV1. KAPP is thought to act as a negative regulator of CLV1 by dephosphorylating it.

Another important gene in plant meristem maintenance is *WUSCHEL* (shortened to *WUS*), which is a target of CLV signalling. *WUS* is expressed in the cells below the stem cells of the meristem and its presence prevents the differentiation of the stem cells. CLV1 acts to promote cellular differentiation by repressing *WUS* activity outside of the central zone containing the stem cells. *STM* also acts to prevent the differentiation of stem cells by repressing the expression of Myb genes that are involved in cellular differentiation.

Root Apical Meristems

Unlike the shoot apical meristem, the root apical meristem produces cells in two dimensions. It is covered by the root cap, which protects the apical meristem from the rocks, dirt and pathogens. Cells are continuously sloughed off the outer surface of the root cap.

The centre of the root apical meristem is occupied by a quiescent centre, which has low mitotic activity. Evidence suggests the quiescent centre does function as the zone of initials. Infrequent division of initial cells in the quiescent centre is the source of cells for the root apical meristem. These initial cells and tissue patterns become established in the embryo in the case of the primary root, and in the new lateral meristems in the case of secondary roots.

Intercalary meristem

In angiosperms, intercalary meristems occur only in monocot (in particular, grass) stems at the base of nodes and leaf blades. Horsetails also exhibit intercalary growth. Intercalary meristems are capable of cell division and allow for rapid growth and regrowth of many monocots. Intercalary meristems at the nodes of bamboo allow for rapid stem elongation, while those at the base of most grass leaf blades allow damaged leaves to rapidly regrow. This leaf regrowth in grasses evolved in response to damage by grazing herbivores, but is more familiar to us in response to lawnmowers.

Floral meristem

When plants begin the developmental process known as flowering, the shoot apical meristem is transformed into an influorescence meristem, which goes on to produce the floral meristem, which produces the familiar sepals, petals, stamens, and carpels of the flower.

In contrast to vegetative apical meristems and some exfluorescence meristems, floral meristems are responsible for determinate growth, the limited growth of the flower to a particular size and form. The transition from shoot meristem to floral meristem requires floral meristem identity genes, that both specify the floral organs and cause the termination of the production of stem cells. *AGAMOUS* (*AG*) is a floral homeotic gene required for floral meristem termination and necessary for proper development of the stamens and carpels.

AG is necessary to prevent the conversion of floral meristems to influorescence shoot meristems, but is not involved in the transition from shoot to floral meristem. AG is turned on by the floral meristem identity gene *LEAFY* (*LFY*) and *WUS* and is restricted to the centre of the floral meristem or the inner two whorls. This way floral identity and region specificity is achieved. WUS activates AG by binding to a consensus sequence in the AG's second intron and LFY binds to adjacent recognition sites. Once AG is activated it represses expression of WUS leading to the termination of the meristem.

Through the years, scientists have manipulated floral meristems for economics reasons. An example is the mutant tobacco plant

"Maryland Mammoth." In 1936, the department of agriculture of Switzerland performed several scientific tests with this plant. "Maryland Mammoth" is peculiar in this sense that it grows much faster than other tobacco plants.

Apical Dominance

Apical dominance is phenomenon where one meristem prevents or inhibits the growth of other meristems. As a result the plant will have one clearly defined main trunk. For example, in trees, the tip of the main trunk bears the dominant meristem. Therefore, the tip of the trunk grows rapidly and is not shadowed by branches. If the dominant meristem is cut off, one or more branch tips will assume dominance. The branch will start growing faster and the new growth will be vertical. Over the years, the branch may begin to look more and more like an extension of the main trunk. Often several branches will exhibit this behaviour after the removal of apical meristem, leading to a bushy growth.

The mechanism of apical dominance is based on the plant hormone auxin. It is produced in the apical meristem and transported towards the roots in the cambium. If apical dominance is complete, it prevents any branches from forming as long as apical meristem is active. If the dominance is incomplete, side branches will develop.

Recent investigations into apical dominance and the control of branching have revealed a new plant hormone family termed strigolactones. These compounds were previously known to be involved in seed germination and communication with mycorrhizal fungi and are now shown to be involved in inhibition of branching.

Diversity in Meristem Architectures

Is the mechanism of being *indeterminate* conserved in the SAM's of the plant world? The SAM contains a population of stem cells that also produce the lateral meristems while the stem elongates. It turns out that the mechanism of regulation of the stem cell number might indeed be evolutionarily conserved. The *CLAVATA* gene *CLV2* responsible for maintaining the stem cell population in *Arabidopsis thaliana* is very closely related to the Maize gene *FASCIATED EAR 2(FEA2)* also involved in the same function. Similarly, in Rice, the *FON1-FON2* system seems to bear a close relationship with the CLV signaling system in *Arabidopsis thaliana*. These studies suggest that the regulation of stem cell number, identity and differentiation might be an evolutionarily conserved mechanism in monocots, if not in angiosperms. Rice also contains another genetic system distinct from *FON1-FON2*, that is involved in regulating stem cell number. This example underlines the innovation that goes about in the living world all the time.

Role of the KNOX-family Genes

Genetic screens have identified genes belonging to the KNOX family in this function. These genes essentially maintain the stem cells in an undifferentiated state. The KNOX family has undergone quite a bit of evolutionary diversification, while keeping the overall mechanism more or less similar. Members of the KNOX family have been found in plants as diverse as Arabidopsis thaliana, rice, barley and tomato. KNOX-like genes are also present in some algae, mosses, ferns and gymnosperms.

Misexpression of these genes leads to formation of interesting morphological features. For example, among members of *Antirrhinae*, only the species of genus *Antirrhinum* lack a structure called spur in the floral region. A spur is considered an evolutionary innovation because it defines pollinator specificity and attraction. Researchers carried out transposon mutagenesis in *Antirrhinum majus*, and saw that some insertions led to formation of spurs that were very similar to the other members of *Antirrhinae*, indicating that the loss of spur in wild *Antirrhinum majus* populations could probably be an evolutionary innovation.

The KNOX family has also been implicated in leaf shape evolution. One study looked at the pattern of KNOX gene expression in *A. thaliana*, that has simple leaves and *Cardamine hirsuta*, a plant having complex leaves. In *A. thaliana*, the KNOX genes are completely turned off in leaves, but in *C.hirsuta*, the expression continued, generating complex leaves. Also, it has been proposed that the mechanism of KNOX gene action is conserved across all vascular plants, because there is a tight correlation between KNOX expression and a complex leaf morphology.

Primary Meristems

Apical meristems may differentiate into three kinds of primary meristem:

- Protoderm - lies around the outside of the stem and develops into the epidermis.
- Ground meristem develops into the Cortex and the pith. It produces the cork cambium, another secondary meristem.

- Procambium - lies just inside of the protoderm and develops into primary xylem and primary phloem. It also produces the vascular cambium, a secondary meristem.

These meristems are responsible for primary growth, or an increase in length or height, which were discovered by scientist Joseph D. Carr of North Carolina in 1943.

Secondary Meristems

There are two types of secondary meristems, these are also called the *lateral meristems* because they surround the established stem of a plant and cause it to grow laterally (i.e., larger in diameter).

- Vascular cambium, which produces secondary xylem and secondary phloem. This is a process that may continue throughout the life of the plant. This is what gives rise to wood in plants. Such plants are called arborescent. This does not occur in plants that do not go through secondary growth (known as herbaceous plants).
- Cork cambium, which gives rise to the periderm, which replaces the epidermis.

Indeterminate Growth of Meristems

Though each plant grows according to a certain set of rules, each new root and shoot meristem can go on growing for as long as it is alive. In many plants, meristematic growth is potentially indeterminate, making the overall shape of the plant not determinate in advance. This is the primary growth. Primary growth leads to lengthening of the plant body and organ formation. All plant organs arise ultimately from cell divisions in the apical meristems, followed by cell expansion and differentiation. Primary growth gives rise to the apical part of many plants.

The growth of nitrogen fixing nodules on legume plants such as soybean and pea is either determinate or indeterminate. Thus, soybean (or bean and Lotus japonicus) produce determinate nodules (spherical), with a branched vascular system surrounding the central infected zone. Often, Rhizobium infected cells have only small vacuoles. In contrast, nodules on pea, clovers, and Medicago truncatula are indeterminate, to maintain (at least for some time) an active meristem that yields new cells for Rhizobium infection. Thus zones of maturity exist in the nodule. Infected cells usually possess a large vacuole. The plant vascular system is branched and peripheral.

Cloning

Under appropriate conditions, each shoot meristem can develop into a complete new plant or clone. Such new plants can be grown from shoot cuttings that contain an apical meristem. Root apical meristems are not readily cloned, however. This cloning is called asexual reproduction or vegetative reproduction and is widely practiced in horticulture to mass-produce plants of a desirable genotype. This process is also known as mericloning.

Propagating through cuttings is another form of vegetative propagation that initiates root or shoot production from secondary meristematic cambial cells. This explains why basal 'wounding' of shoot-borne cuttings often aids root formation.

Bibliography

Acharya, Deepak; Anshu, Shrivastava: *Indigenous Herbal Medicines: Tribal Formulations and Traditional Herbal Practices*, Aavishkar Publishers, Jaipur, India, 2008.

Ackerknecht, Erwin H.: *Medicine and Ethnology*. Johns Hopkins Press, Baltimore, 1971.

Anderson, Edgar: *Plants, Man and Life*. U. Calif. Press, Berkeley, 1967.

Barrows, D. P.: *The Ethno-Botany of the Coahuilla Indians of Southern California*. Univ. Chicago Press, 1900.

Bellamy, David: *Bellamy on Botany*, Edbury Publishing - BBC Books, London, 1972.

Capon, Brian: *Botany for Gardeners*, Timber Publishing, Portland, OR, 2005.

Castetter, E. F., W. H. Bell, & A. R. Grove: *Albuquerque*, Univ. N. Mex. Press, Southwest, 1938.

Culberson, C. F. 1969. Chemical and Botanical Guide to Lichen Products. Univ. of North Carolina Press, Chapel Hill.

Dallal, Ahmad: *Islam, Science, and the Challenge of History*, Yale University Press, New Haven, CT, 2010.

Gantt, Elisabeth: *Handbook of Phycological Methods. Developmental and Cytological Methods*. Cambridge University Press, Cambridge, England, 1980.

Halle, Francis: *In Praise of Plants*, Timber Publishing, Portland, OR, 2002.

Hassrick, R. B.: *The Sioux. Norman*, Univ. of Oklahoma Press, Oklahoma, 1964.

Hellebust, J. A. and J. S. Craigie: *Handbook of Phycological Methods. Physiological and Biochemical Methods*. Cambridge University Press, Cambridge, England, 1978.

Hussey, P. B.: *A taxonomic list of some plants of economic importance*. The Science Press Printing Co., Lancaster, Pa, 1939.

Iyer, Meena: *Faith & Philosophy of Zoroastrianism*, Kalpaz Publications, Delhi, India, 2009.

Jeanness, Diamond: *The People of the Twilight*. V. Chicago Press, Chicago, 1959.

King, John: *Reaching for the Sun: How Plants Work*. Cambridge University Press, Cambridge, 1997.

Lloyd, John U.: *Origin and History of all the Pharmacopoeial Vegetable Drugs, Chemicals, and Preparations*, The Caxton Press, Cincinnati, 1998.

Martin, A. C. & W. D. Barkley: *Seed Identification Manual*. U. Cal. Press, Berkeley, 1961.

Pakenham, Thomas: *Remarkable Trees of the World*, W. W. Norton, London, 2002.

Pollan, Michael: *The Botany of Desire: A Plant's-eye View of the World*, Public Broadcasting System (TV), Random House, New York, 2001.

Rayback, R. J.: *The Indian*, Syracuse Univ. Press. Syracuse, NY, 1966.

Rutsch, E. S.: *Smoking technology of the aborigines of the Iroquois area*, Associated Univ. Press, NY,1973.

Thomas, Barry A.: *The Evolution of Plants and Flowers*, St. Martin's Press, New York, 1981.

Walker, David: *Energy, Plants and Man*, Oxygraphics Ltd., Sheffield, England, 1999.

Will, G. F., & G. E. Hyde: *Corn among the Indians of the upper Missouri*. Univ. Neb. Press, Lincoln, 1917.

Yaniv, Zohara; Bachrach, Uriel: *Handbook of Medicinal Plants*, Haworth Press, Binghampton, NY, 2005.

Young, James Harvey: *The Toadstool Millionaires: A social historyof patent medicines in America before Federal regulation*. Princeton Univ. Press, Princeton, 1961.

Index

H

I

J

L

M

N

P

R

S

T

V

W

Z

❑❑❑